COMMITTEE ON CARBON MONOXIDE EPISODES IN METEOROLOGICAL AND TOPOGRAPHICAL PROBLEM AREAS

Members

ARMISTEAD G. RUSSELL *(Chair)*, Georgia Institute of Technology, Atlanta
ROGER ATKINSON, University of California, Riverside
SUE ANN BOWLING, University of Fairbanks (Retired), Fairbanks, AK
STEVEN D. COLOME, University of California, Los Angeles
NAIHUA DUAN, University of California, Los Angeles
GERALD GALLAGHER, J Gallagher and Associates, Inc., Englewood, CO
RANDALL L. GUENSLER, Georgia Institute of Technology, Atlanta
SUSAN L. HANDY, University of California, Davis
SIMONE HOCHGREB, University of Cambridge, England
SANDRA N. MOHR, Consultant, Gillette, New Jersey
ROGER A. PIELKE SR., Colorado State University, Fort Collins
KARL J. SPRINGER, Southwest Research Institute (Retired), San Antonio, TX
ROGER WAYSON, University of Central Florida, Orlando

Project Staff

K. JOHN HOLMES, Senior Staff Officer
RAYMOND WASSEL, Senior Program Director
NANCY HUMPHREY, Senior Staff Officer
CHAD TOLMAN, Staff Officer
LAURIE GELLER, Senior Staff Officer
AMANDA STAUDT, Staff Officer
KELLY CLARK, Editor
MIRSADA KARALIC-LONCAREVIC, Research Assistant
RAMYA CHARI, Research Assistant
EMILY BRADY, Senior Project Assistant

Sponsor

U.S. ENVIRONMENTAL PROTECTION AGENCY

BOARD ON ENVIRONMENTAL STUDIES AND TOXICOLOGY

Members

GORDON ORIANS *(Chair)*, University of Washington, Seattle
JOHN DOULL *(Vice Chair)*, University of Kansas Medical Center, Kansas City
DAVID ALLEN, University of Texas, Austin
THOMAS BURKE, Johns Hopkins University, Baltimore, MD
JUDITH C. CHOW, Desert Research Institute, Reno, NV
CHRISTOPHER B. FIELD, Carnegie Institute of Washington, Stanford, CA
WILLIAM H. GLAZE, Oregon Health and Science University, Beaverton
SHERRI W. GOODMAN, Center for Naval Analyses, Alexandria, VA
DANIEL S. GREENBAUM, Health Effects Institute, Cambridge, MA
ROGENE HENDERSON, Lovelace Respiratory Research Institute, Albuquerque, NM
CAROL HENRY, American Chemistry Council, Arlington, VA
ROBERT HUGGETT, Michigan State University, East Lansing
BARRY L. JOHNSON Emory University, Atlanta, GA
JAMES H. JOHNSON, Howard University, Washington, DC
JAMES A. MACMAHON, Utah State University, Logan
PATRICK V. O'BRIEN, Chevron Research and Technology, Richmond, CA
DOROTHY E. PATTON, International Life Sciences Institute, Washington, DC
ANN POWERS, Pace University School of Law, White Plains, NY
LOUISE M. RYAN, Harvard University, Boston, MA
JONATHAN M. SAMET, Johns Hopkins University, Baltimore, MD
KIRK SMITH, University of California, Berkeley
LISA SPEER, Natural Resources Defense Council, New York, NY
G. DAVID TILMAN, University of Minnesota, St. Paul
CHRIS G. WHIPPLE, Environ Incorporated, Emeryville, CA
LAUREN A. ZEISE, California Environmental Protection Agency, Oakland, CA

Senior Staff

JAMES J. REISA, Director
DAVID J. POLICANSKY, Associate Director
RAYMOND A. WASSEL, Senior Program Director for Environmental Sciences and Engineering
KULBIR BAKSHI, Program Director for the Committee on Toxicology
ROBERTA M. WEDGE, Program Director for Risk Analysis
K. JOHN HOLMES, Senior Staff Officer
SUSAN N.J. MARTEL, Senior Staff Officer
SUZANNE VAN DRUNICK, Senior Staff Officer
EILEEN N. ABT, Senior Staff Officer
ELLEN K. MANTUS, Senior Staff Officer
RUTH E. CROSSGROVE, Managing Editor

MANAGING **CARBON MONOXIDE POLLUTION** IN METEOROLOGICAL AND TOPOGRAPHICAL PROBLEM AREAS

Committee on Carbon Monoxide Episodes in
Meteorological and Topographical Problem Areas

Board on Environmental Studies and Toxicology
Board on Atmospheric Sciences and Climate
Division on Earth and Life Studies
Transportation Research Board

NATIONAL RESEARCH COUNCIL
OF THE NATIONAL ACADEMIES

THE NATIONAL ACADEMIES PRESS
Washington, D.C.
www.nap.edu

THE NATIONAL ACADEMIES PRESS   500 Fifth Street, NW   Washington, D.C. 20001

NOTICE: The project that is the subject of this report was approved by the Governing Board of the National Research Council, whose members are drawn from the councils of the National Academy of Sciences, the National Academy of Engineering, and the Institute of Medicine. The members of the committee responsible for the report were chosen for their special competences and with regard for appropriate balance.

This project was supported by Contract No. X-82880601-0 between the National Academy of Sciences and the U.S. Environmental Protection Agency. Any opinions, findings, conclusions, or recommendations expressed in this publication are those of the author(s) and do not necessarily reflect the view of the organizations or agencies that provided support for this project.

International Standard Book Number 0-309-08923-9 (Book)

International Standard Book Number 0-309-50884-3 (PDF)

Additional copies of this report are available from:

The National Academies Press
500 Fifth Street, NW
Box 285
Washington, DC 20055

800-624-6242
202-334-3313 (in the Washington metropolitan area)
http://www.nap.edu

Copyright 2003 by the National Academy of Sciences. All rights reserved.

Printed in the United States of America

Cover: photograph by Nick Wheeler/CORBIS

THE NATIONAL ACADEMIES
Advisers to the Nation on Science, Engineering, and Medicine

The **National Academy of Sciences** is a private, nonprofit, self-perpetuating society of distinguished scholars engaged in scientific and engineering research, dedicated to the furtherance of science and technology and to their use for the general welfare. Upon the authority of the charter granted to it by the Congress in 1863, the Academy has a mandate that requires it to advise the federal government on scientific and technical matters. Dr. Bruce M. Alberts is president of the National Academy of Sciences.

The **National Academy of Engineering** was established in 1964, under the charter of the National Academy of Sciences, as a parallel organization of outstanding engineers. It is autonomous in its administration and in the selection of its members, sharing with the National Academy of Sciences the responsibility for advising the federal government. The National Academy of Engineering also sponsors engineering programs aimed at meeting national needs, encourages education and research, and recognizes the superior achievements of engineers. Dr. Wm. A. Wulf is president of the National Academy of Engineering.

The **Institute of Medicine** was established in 1970 by the National Academy of Sciences to secure the services of eminent members of appropriate professions in the examination of policy matters pertaining to the health of the public. The Institute acts under the responsibility given to the National Academy of Sciences by its congressional charter to be an adviser to the federal government and, upon its own initiative, to identify issues of medical care, research, and education. Dr. Harvey V. Fineberg is president of the Institute of Medicine.

The **National Research Council** was organized by the National Academy of Sciences in 1916 to associate the broad community of science and technology with the Academy's purposes of furthering knowledge and advising the federal government. Functioning in accordance with general policies determined by the Academy, the Council has become the principal operating agency of both the National Academy of Sciences and the National Academy of Engineering in providing services to the government, the public, and the scientific and engineering communities. The Council is administered jointly by both Academies and the Institute of Medicine. Dr. Bruce M. Alberts and Dr. Wm. A. Wulf are chair and vice chair, respectively, of the National Research Council.

www.national-academies.org

TRANSPORTATION RESEARCH BOARD EXECUTIVE COMMITTEE

Members

GENEVIEVE GIULIANO (*Chair*), University of Southern California, Los Angeles
MICHAEL S. TOWNES (*Vice Chair*), Hampton Roads Transit, Virginia
ROBERT E. SKINNER, JR. *(Executive Director)*, National Research Council, Washington, DC
MICHAEL W BEHRENS, Texas Department of Transportation, Austin
JOSEPH H. BOARDMAN, New York State Department of Transportation, Albany
SARAH C. CAMPBELL, TransManagement Inc., Washington DC
E. DEAN CARLSON, Independent Consultant, Topeka, KS
JOANNE CASEY, Intermodal Association of North America, Greenbelt, MD
JAMES C. CODELL III, Kentucky Transportation Cabinet, Frankfort
JOHN L. CRAIG, Nebraska Department of Roads, Lincoln
BERNARD S. GROSECLOSE JR., South Carolina State Ports Authority, Charleston
SUSAN HANSON, Clark University, Worcester, MA
LESTER A. HOEL, University of Virginia, Charlottesville
HENRY L. HUNGERBEELER, Missouri Department of Transportation, Jefferson City
ADIB K. KANAFANI, University of California, Berkeley
RONALD F. KIRBY, Metropolitan Washington Council of Governments, Washington DC
HERBERT S. LEVINSON, Herbert S. Levinson Transportation Cons., New Haven, CT
MICHAEL D. MEYER, Georgia Institute of Technology, Atlanta
JEFF P. MORALES, California Department of Transportation, Sacramento
KAM K. MOVASSAGHI, Louisiana Department of Transportation & Development, Baton Rouge
CAROL A. MURRAY, New Hampshire Department of Transportation, Concord
DAVID Z. PLAVIN, Airports Council International of North America, Washington DC
JOHN H. REBENSDORF, Union Pacific Railroad Company, Omaha
CATHERINE L. ROSS, Consultant, Atlanta, GA
JOHN M. SAMUELS, Norfolk Southern Corporation, Norfolk, VA
PAUL P. SKOUTELAS, Port Authority of Allegheny County, Pittsburgh, PA
MICHAEL W. WICKHAM, Roadway Express, Inc., Akron, OH

BOARD ON ATMOSPHERIC SCIENCES AND CLIMATE

Members

ERIC J. BARRON (*Chair*), Pennsylvania State University, University Park
RAYMOND J. BAN, The Weather Channel, Inc., Atlanta, GA
ROBERT C. BEARDSLEY, Woods Hole Oceanographic Institution, Woods Hole, MA
ROSINA M. BIERBAUM, The University of Michigan, Ann Arbor
HOWARD B. BLUESTEIN, University of Oklahoma, Norman
RAFAEL L. BRAS, Massachusetts Institute of Technology, Cambridge
STEVEN F. CLIFFORD, University of Colorado, Boulder
CASSANDRA G. FESEN, Dartmouth College, Hanover, NH
GEORGE L. FREDERICK, Vaisala Meteorological Systems, Inc., Boulder, CO
JUDITH L. LEAN, Naval Research Laboratory, Washington, DC
MARGARET A. LEMONE, National Center for Atmospheric Research, Boulder, CO
MARIO J. MOLINA, Massachusetts Institute of Technology, Cambridge
MICHAEL J. PRATHER, University of California, Irvine
WILLIAM J. RANDEL, National Center for Atmospheric Research, Boulder, CO
RICHARD D. ROSEN, Atmospheric & Environmental Research, Inc., Lexington, MA
THOMAS F. TASCIONE, Sterling Software, Inc., Bellevue, NE
JOHN C. WYNGAARD, Pennsylvania State University, University Park

OTHER REPORTS OF THE
BOARD ON ENVIRONMENTAL STUDIES AND TOXICOLOGY

Cumulative Environmental Effects of Alaska North Slope Oil and Gas Development (2003)
Estimating the Public Health Benefits of Proposed Air Pollution Regulations (2002)
Biosolids Applied to Land: Advancing Standards and Practices (2002)
Ecological Dynamics on Yellowstone's Northern Range (2002)
The Airliner Cabin Environment and Health of Passengers and Crew (2002)
Arsenic in Drinking Water: 2001 Update (2001)
Evaluating Vehicle Emissions Inspection and Maintenance Programs (2001)
Compensating for Wetland Losses Under the Clean Water Act (2001)
A Risk-Management Strategy for PCB-Contaminated Sediments (2001)
Acute Exposure Guideline Levels for Selected Airborne Chemicals (3 volumes, 2000-2003)
Toxicological Effects of Methylmercury (2000)
Strengthening Science at the U.S. Environmental Protection Agency (2000)
Scientific Frontiers in Developmental Toxicology and Risk Assessment (2000)
Ecological Indicators for the Nation (2000)
Modeling Mobile-Source Emissions (2000)
Waste Incineration and Public Health (1999)
Hormonally Active Agents in the Environment (1999)
Research Priorities for Airborne Particulate Matter (4 volumes, 1998-2003)
Ozone-Forming Potential of Reformulated Gasoline (1999)
Arsenic in Drinking Water (1999)
The National Research Council's Committee on Toxicology: The First 50 Years (1997)
Carcinogens and Anticarcinogens in the Human Diet (1996)
Upstream: Salmon and Society in the Pacific Northwest (1996)
Science and the Endangered Species Act (1995)
Wetlands: Characteristics and Boundaries (1995)
Biologic Markers (5 volumes, 1989-1995)
Review of EPA's Environmental Monitoring and Assessment Program (3 volumes, 1994-1995)
Science and Judgment in Risk Assessment (1994)
Pesticides in the Diets of Infants and Children (1993)
Dolphins and the Tuna Industry (1992)
Science and the National Parks (1992)
Human Exposure Assessment for Airborne Pollutants (1991)
Rethinking the Ozone Problem in Urban and Regional Air Pollution (1991)
Decline of the Sea Turtles (1990)

Copies of these reports may be ordered from the National Academy Press
(800) 624-6242 or (202) 334-3313
www.nap.edu

Acknowledgment of Review Participants

This report has been reviewed in draft form by individuals chosen for their diverse perspectives and technical expertise, in accordance with procedures approved by the National Research Council's (NRC's) Report Review Committee. The purpose of this independent review is to provide candid and critical comments that will assist the institution in making its published report as sound as possible and to ensure that the report meets institutional standards for objectivity, evidence, and responsiveness to the study charge. The review comments and draft manuscript remain confidential to protect the integrity of the deliberative process. We wish to thank the following individuals for their review of this report:

Richard Baker, Ford Motor Company
Lenora Bohren, Colorado State University
Russell Dickerson, University of Maryland
Robert Dulla, Sierra Research, Inc.
Peter Flachsbart, University of Hawaii at Manoa
Bernard D. Goldstein, University of Pittsburgh
R. Michael Hardesty, National Oceanic and Atmospheric Administration
Roland Hwang, Natural Resources Defense Council
Jana Milford, University of Colorado at Boulder
William Neff, National Oceanic and Atmospheric Administration
Robert F. Klausmeier, de la Torre Klausmeier Consulting, Inc.

Andrew Sessler, Lawrence Berkeley National Laboratory
Michael Walsh, Consultant

Although the reviewers listed above have provided many constructive comments and suggestions, they were not asked to endorse the conclusions or recommendations, nor did they see the final draft of the report before its release. The review of this report was overseen by William Chameides, Georgia Institute of Technology, and Robert Sawyer, University of California, Berkeley. Appointed by the NRC, they were responsible for making certain that an independent examination of this report was carried out in accordance with institutional procedures and that all review comments were carefully considered. Responsibility for the final content of this report rests entirely with the authoring committee and the institution.

Preface

Carbon monoxide (CO) is a toxic air pollutant emitted largely from motor vehicles. It is a colorless, odorless, tasteless gas that can produce serious adverse health effects. Breathing CO at high concentrations leads to carboxyhemoglobin poisoning (reduced oxygen transport by hemoglobin), which can lead to impaired reaction timing, headaches, lightheadedness, nausea, coma, and, at high enough concentrations and long enough exposure, death. At lower concentrations that can occur in the ambient environment, the effects of CO exposure include increased risk of chest pain and hospitalization for persons with coronary artery disease. Because of the adverse health effects associated with this pollutant, the U.S. Environmental Protection Agency (EPA), as directed by the Clean Air Act, established the health-based National Ambient Air Quality Standards (NAAQS) for CO in 1971.

Reducing CO pollution has been one of the greatest success stories in emissions control. Over the past three decades improved motor-vehicle emissions controls have greatly reduced ambient CO concentrations. Most areas that were originally designated as nonattainment areas have come into compliance with the NAAQS for CO. However, certain locations continue to occasionally experience high concentrations of CO. These locations tend to have topographical and meteorological characteristics that exacerbate pollution. Compliance with the health-based NAAQS for CO has proved difficult under those circumstances. In response to the challenges posed for certain areas by having to come into compliance with the

NAAQS for CO, Congress requested that the National Research Council investigate the characteristics of CO in areas with meteorological and topographical handicaps. In an interim report released in May of 2002, the committee addressed this issue for Fairbanks, Alaska.

Many people assisted the committee by providing information related to issues addressed in this report. We gratefully acknowledge Steven Albu, California Air Resources Board; Joseph Cassmassi, South Coast Air Quality Management District; Bart Croes, California Air Resources Board; Greg Dana, Alliance of Automobile Manufacturers; Laurence Elmore, EPA; Robert Gibbons, University of Illinois at Chicago; Douglas Lawson, National Renewable Energy Laboratory; William Neff, National Oceanic and Atmospheric Administration; Dennis Ransel, Clark County, Nevada; Patrick Reddy, Colorado Department of Public Health and the Environment; Beate Ritz, University of California, Los Angeles; Shannon Therriault, Missoula City/County Health Department.

I am also grateful for the assistance of the National Research Council staff in the preparation of this report. K. John Holmes greatly assisted the committee in his role as project director. The committee also acknowledges Raymond A. Wassel, senior program director for environmental sciences and engineering in the Board on Environmental Studies and Toxicology (BEST). We thank the other staff members who contributed to this report, including Warren Muir, executive director of the Division on Earth and Life Studies; James J. Reisa, director of BEST; Nancy Humphrey, senior staff officer with the Transportation Research Board; Laurie Geller, senior staff officer with the Board on Atmospheric Sciences and Climate (BASC); Amanda Staudt, staff officer with BASC; Chad Tolman, staff officer with BEST (retired); Kelly Clark, assistant editor with BEST; Mirsada Karalic-Loncarevic, research assistant with BEST; Ramya Chari, research assistant with BEST; and Emily Brady, senior project assistant with BEST.

Finally, I would like to thank all the members of the committee for their expertise and dedicated effort throughout the study.

Armistead Russell, Ph.D.
Chair, Committee on Carbon Monoxide Episodes in Meteorological and Topographical Problem Areas

Contents

Summary 1

1 Ambient Carbon Monoxide Pollution in the United States 16
   Introduction, *16*
   Study Background and Charge, *17*
   Summary of Interim Report, *21*
   The Committee's Approach To Its Charge, *22*
   Report Contents, *23*
   National Regulatory Setting for Ambient CO, *23*
   Areas With Recent Exceedances of the CO Standard, *27*
   Sources of CO Emissions, *31*
   Health Effects of CO, *39*
   Relationship of CO to Other Air Pollutants, *51*
   Equity Considerations in the Spatial Distribution of
      Ambient CO, *65*

2 Contributions of Topography, Meteorology, and
   Human Activity to Carbon Monoxide Concentrations 72
   Introduction, *72*
   Meteorology and Topography, *74*
   Temporal Patterns of CO Concentrations, *82*
   Vulnerability to Future Exceedances, *88*
   Illustrative Examples, *94*

3 Management of Carbon Monoxide Air Quality.................................100
   Emissions Control Programs, *100*
   Monitoring, Models, and Inventories, *129*

4 The Future of Carbon Monoxide Air Quality Management...............*149*
   Exposures of Concern in the Future, *150*
   Future CO Management Issues, *151*
   Integrating CO Control Into the Overall Air Quality
     Management System, *156*

References..*160*

Glossary..*178*

Appendix A. Biographical Information on the Committee on
   Carbon Monoxide Episodes in Meteorological and
   Topographical Problem Areas..*189*

Appendix B. Abbreviations and Names Used for Classifying
   Organic Compounds..*193*

Appendix C. A Simple Box Model with Recirculation........................*194*

Managing Carbon Monoxide Pollution in Meteorological and Topographical Problem Areas

Summary

A primary objective of air quality management in the United States has been to reduce human exposure to carbon monoxide (CO) and other pollutants produced from incomplete combustion. Elevated ambient CO concentrations are due mainly to incomplete combustion of gasoline by light-duty vehicles, such as passenger cars and pickup trucks.

CO controls are working. Problems with ambient CO were widespread when automobile emissions regulations began in the 1960s. When the health-based National Ambient Air Quality Standards (NAAQS) for CO were promulgated in 1971, more than 90% of ambient monitors indicated violations.[1] Since then, motor-vehicle emissions controls have greatly reduced ambient CO concentrations. Over the last 5 years, the number of monitors showing CO violations has fallen to only a few, and the monitors that continue to show violations do so much less frequently. For example, Denver, Colorado, which had a persistent CO problem and registered as many as 200 days with violations in the 1960s, has not had a violation since 1995. Fairbanks, Alaska, reduced the number of days with violations from

---

[1]The standards for ambient air concentrations of CO were set at 9 parts per million (ppm) for an 8-hour average and 35 ppm for a 1-hour average. These standards were set to protect public health with "an adequate margin of safety," as specified in the Clean Air Act. A violation of an NAAQS occurs on the second exceedance and all subsequent exceedances of the standard in a calendar year. Only the 8-hour standard of 9 ppm has been exceeded recently in a few locations in the country.

well over 100 per year in the early 1970s to zero over the last 2 years. Thus, CO regulation has been one of the greatest success stories in air pollution control, reducing the problem, once widespread, to a few difficult areas. As a result, the focus of U.S. air quality management has shifted to characterizing and controlling other pollutants, such as tropospheric ozone, fine particulate matter ($PM_{2.5}$),[2] and air toxics.

However, some locations, such as Anchorage, Alaska, and Lynwood and Calexico, California, continue to be susceptible to occasional violations of the NAAQS for CO. These areas are typically subject to problematic meteorological and topographical conditions that produce severe atmospheric inversions in winter.[3] Although CO emissions from light-duty vehicles are projected to decrease in the future, atmospheric inversions and low windspeeds prevalent in some locations during winter are extremely effective in trapping the products of incomplete combustion, including CO, emitted at ground level. For example, Fairbanks, Alaska, is subject to extreme atmospheric inversions, at times experiencing inversion strengths as much as 30°C per 100 meters of altitude. In addition, Fairbanks is situated in a three-sided bowl, surrounded by the Yukon-Tanana uplands, which can inhibit air circulation. Although it is not heavily populated and has no major air-pollution producing industries, Fairbanks's meteorological and topographical characteristics make the city susceptible to high ambient CO concentrations in winter.

The continuing vulnerability of a few locations to high CO concentrations prompted Congress, in its fiscal 2001 appropriations report for the U.S. Environmental Protection Agency (EPA), to ask the National Academy of Sciences to study CO episodes in meteorological and topographical problem areas. The study was requested to address potential approaches to predicting, assessing, and managing episodes of high CO concentrations in such areas. In particular, the committee was to address:

---

[2]$PM_{2.5}$ is a subset of particulate matter that includes particles with an aerodynamic equivalent diameter less than or equal to a nominal 2.5 micrometers.

[3]Inversions occur when the temperature of the atmosphere increases with altitude. Combined with low windspeeds, this prevents air circulation, because colder air is trapped near the ground by the warmer air above. A temperature increase of several degrees celcius per 100 meters is considered a strong inversion. The standard lapse rate for the troposphere is a decrease of about 6.5°C per kilometer (or about 3.6°F per 1,000 feet).

- Types of emission sources and operating conditions that contribute most to episodes of high ambient CO.
- Scientific bases of current and potential additional approaches for developing and implementing plans to manage CO air quality, including the possibility of new catalyst technology, alternative fuels, and cold-start technology, as well as traffic and other management programs for motor-vehicle sources. Control of stationary-source contributions to CO air quality also was to be considered.
- The effectiveness of CO emissions control programs, including comparisons among areas with and without unusual topographical or meteorological conditions.
- Relationships between monitored episodes of high ambient CO concentrations and personal human exposure.
- The public-health impact of such episodes.
- Statistically robust alternative methods to assist in tracking progress in reducing CO that bear a relation to the CO concentrations considered harmful to human health.

In response, the National Research Council convened the Committee on Carbon Monoxide Episodes in Meteorological and Topographical Problem Areas, which prepared this report. Fairbanks, Alaska, was identified as the subject for a case study in an interim report, which was completed in 2002. The following report is the final report requested by Congress.

FINDINGS AND RECOMMENDATIONS

Vulnerability to Future Violations

Findings

Because of a number of factors—including differences in topography and temporal variability of local meteorology and emissions rates—some areas are especially vulnerable to violations of the 8-hour NAAQS for CO. In geographical areas that have achieved attainment of the NAAQS, it might still be possible for ambient concentrations of CO to sporadically exceed the standard under unfavorable conditions, such as strong winter inversions. This vulnerability, defined as the probability for violation in a future year, depends on both the current CO levels and the variability of air quality indicators. An area in attainment might still be substantially vulnerable if the variability of its air quality is high.

There is evidence that local meteorological conditions conducive to high CO concentrations are sometimes associated with large-scale meteorological and climatological phenomena. For example, all recent exceedances of the NAAQS for CO in Fairbanks have occurred with a low-pressure system in the Gulf of Alaska with cyclonic flow extending over Fairbanks. Although the role that this low-pressure system plays is unclear, it might produce warm winds aloft that reinforce inversions near the ground. In Denver, Colorado, the presence of long-term snow cover and light winds can produce conditions conducive to CO buildup in the ambient air. Snow cover diminishes ground-level solar heating, intensifying inversions, and light winds reduce horizontal dispersion. However, over the past decade, Denver has not experienced the combination of these meteorological factors, reducing the city's susceptibility to high CO concentrations. Changes in the frequency of some large-scale meteorological and climatological events, such as the frequency of low-pressure systems in the Gulf of Alaska, will influence vulnerability to CO violations.

Recommendations

Air quality managers typically recognize whether their region is especially vulnerable to future CO violations as a result of increases in vehicle activity, the spatial and temporal variability of meteorology, and problematic topography. However, in some cases, air quality planning does not encompass the worst-case combinations of emissions and meteorology. Achieving sufficient emissions reductions to account for these conditions is prudent, particularly in areas with high population growth and/or high meteorological variability, to further reduce the risk of violations. In addition, given that the form of the CO standard defines a violation as the second and all subsequent exceedances in a calendar year, regions are susceptible to violating the standard due to extreme meteorological conditions, which contributes to the difficulties that meteorological and topographical problem areas have in reaching and maintaining attainment. It is also important to investigate how large-scale and local meteorological and climatological phenomena can affect the susceptibility of a location to CO buildup in ambient air.

Air quality managers should recognize that their regions might be especially vulnerable to future CO violations because of increases in vehicle activity, spatial and temporal variability of meteorology, and problematic topography. The meteorological conditions assumed in current regulatory air quality planning might not encompass the worst-case conditions. Achieving additional emissions reductions is prudent to further reduce the risk of violations, particularly in areas with high population growth and/or high meteorological variability. It also might be important to investigate how large-scale meteorological and climatological phenomena can affect the susceptibility of a location to CO buildup in ambient air.

Health Effects

Findings

In patients diagnosed with coronary artery disease, CO alone has been shown to exacerbate exercise-induced chest pain (angina) in controlled laboratory experiments. Those studies serve as an important part of the basis for the NAAQS. In addition, epidemiological studies have correlated high CO concentrations with other adverse human health effects, such as heart disease, childhood developmental abnormalities, and fetal loss. Some of these effects have been correlated with ambient CO levels below the NAAQS. However, CO is not produced alone, and epidemiological studies have difficulty separating the effects of CO from those of other pollutants that are often associated with CO (such as benzene, 1,3-butadiene, aldehydes, and various components of $PM_{2.5}$). Though changes in ambient levels of CO may sometimes correlate with health effects other than exercise-induced angina, there is insufficient evidence that CO is the single direct causative agent in the other effects. Thus, reducing CO alone may or may not reduce the incidence of heart disease, childhood developmental abnormalities, and fetal loss.

A significant collateral benefit from reducing CO vehicle emissions standards has been the substantial reduction in accidental deaths due to acute CO poisoning. CO is unique among criteria pollutants[4] because it is relevant to both ambient air quality management and public safety. Expo-

---

[4]Criteria pollutants are air pollutants emitted from numerous or diverse stationary or mobile sources for which NAAQS have been set to protect human health and public welfare. The other criteria pollutants are ozone, particulate matter, sulfur dioxide, nitrogen dioxide, and lead.

sures to mobile-source emissions cross these two areas of public-health management. Using computerized death-certificate data, a recent study by researchers at the Centers for Disease Control and Prevention estimated that over 11,000 deaths from accidental CO poisoning have been avoided over the 1968-1998 time period because of the more stringent vehicle-emissions standards. This collateral benefit is not accounted for in EPA's recent report to Congress on the benefits and costs of the Clean Air Act.

Recommendations

To reduce the potential adverse health effects of CO, the few remaining areas not in attainment need to continue making progress towards meeting and maintaining the CO standard. Public-health issues associated with ambient CO should be emphasized through enhanced public-awareness campaigns. Further study to reveal the effects of CO on the fetus and to separate the effects of CO from its copollutants is encouraged. Also, there should be more toxicology studies of the automobile exhaust mixture.

Management and Control of CO

Management of CO

Findings

To reach attainment, communities vulnerable to exceeding the health-based NAAQS for CO can implement various local measures to complement federal vehicle emissions standards. These include but are not limited to vehicle emissions inspection and maintenance (I/M) programs, the use of cold weather engine-block heaters in vehicles, and the use of low-sulfur gasoline and oxygenated fuels. These measures reduce CO emissions by reducing the number of malfunctioning vehicles (I/M programs), reducing the length of time before a vehicle's emissions control catalyst is fully functional (cold weather engine-block heaters), and improving the efficiency of the vehicle's emissions control catalyst (low-sulfur gasoline). Other measures include mass transit initiatives, traffic management, and bans on wood burning.

Air quality issues have tended to be assessed and regulated independently. As such, CO management frequently occurs in isolation, even though other pollutants have similar emissions sources. The committee recognizes that few areas in the country continue to violate CO standards and that the focus of air quality management in the near future will not be on CO but on attaining new $PM_{2.5}$ and tropospheric ozone air quality standards and reducing air toxics.

Recommendations

Communities with special CO problems should be encouraged to design locally effective programs. Federal and state assistance should be provided to these communities for characterization and implementation of management options. This should include assistance to improve non-road and stationary-source emissions characterizations. Because the CO standards are health-based, all communities need to be diligent in working toward attaining and maintaining the CO standards. In addition, the programs implemented to reduce CO emissions should be reassessed periodically. This reassessment should include evaluation of their impact on CO, as well as other pollutants, and their impact at low temperatures.

CO management should be better integrated into air quality management. Although the focus of air quality management in the near future will be on other air pollution issues, winter inversion conditions not only affect CO buildup but can also be related to higher concentrations of $PM_{2.5}$ and some air toxics. In addition, the primary source of CO, fuel-rich operations of light-duty vehicles, is a major source of other pollutants of concern. The committee therefore recommends that EPA assess the relationship of CO to these other pollutants when the CO criteria are updated.

Federal Tier 2 and Cold-Start Emissions Standards

Findings

Federal new-vehicle emissions standards have been effective in reducing CO emissions, including emissions from vehicles operated in cold climates. Emissions from passenger vehicles have been reduced from their pre-control levels of over 80 grams per mile (g/mi) in the late 1960s to the 3.4 g/mi standard implemented in 1981. Also, progressively lower standards for hydrocarbon emissions, as well as other requirements, have tended to decrease CO emissions levels. Since 1994, new cars have been required to meet a winter cold-start CO limit, which reduces emissions from vehicles started at cold temperatures (20°F).

New Tier 2 and California certification standards[5] are expected to further reduce hydrocarbon and nitrogen oxides (NO_x) emissions. Average CO emissions from vehicles certified to Tier 2 standards are expected to decrease due to the technological improvements in emissions control systems needed to meet the standards, especially the lower hydrocarbon standards and lower CO standards on some light-duty trucks.

In CO problem areas, the decrease in emissions resulting from Federal Tier 2 standards will depend on the mix of vehicles sold and used in these areas. As such, the following uncertainties arise in predicting the continued motor-vehicle emissions reductions in CO problem areas:

- The sales strategy used by manufacturers to comply with average NO_x limits. If manufacturers tend to sell higher-emitting vehicles in CO problem areas, the improvements in CO emissions will not be as large as those predicted based on national averages.[6]
- The impact of the emissions averaging and trading provisions (which allow vehicle manufacturers to generate, trade, and bank emissions credits) on the fleet of vehicles operating in CO problem areas.[7]
- The effects of Tier 2 requirements on CO emissions at temperatures below 20°F.

---

[5]Federal Tier 2 emissions standards will be introduced for passenger cars in model-year 2004 and fully implemented by model-year 2007. The standards will require that vehicle manufacturers meet a fleet average NO_x limit of 0.07 grams per mile (g/mi) along with lower standards for hydrocarbons. California, which is allowed under the Clean Air Act to adopt its own vehicle emissions standards, has already implemented a similar set of emissions limits. Both the Tier 2 and California standards are for vehicles certified at 68-86°F. Since 1994, new cars also have been subject to a cold-start CO standard, which requires cars and most light-duty trucks to meet a CO limit of 10 g/mi on a certification test run at 20°F. The Clean Air Act Amendments of 1990 include a provision for more stringent cold-start standards to be set if needed.

[6]The Tier 2 standard is a sales-weighted fleet-averaged standard. Thus, some vehicles that are sold will have emissions greater than the standard, some less than the standard.

[7]The Tier 2 standard has provisions that allow manufacturers bettering the fleet-averaged standard to generate tradable emissions credits that can be sold to manufacturers who have not met the fleet-averaged standard. Manufacturers can also bank credits for use in later years.

- The ability of EPA's MOBILE emissions rate model to adequately account for the effects of Tier 2 standards on CO emissions rates.[8]

Recommendations

In the absence of compelling evidence, the committee does not recommend tightening the national cold-start standard below 10.0 g/mi or requiring that the 10.0 g/mi standard be met at a lower temperature. However, supplemental emissions testing should be undertaken at temperatures below 20°F to determine to what extent CO emissions systematically increase as ambient temperature decreases. Testing data should be obtained and analyzed at 0°F and 10°F, and should include CO as well as other pollutants (air toxics and PM).

The extent of the anticipated reduction in CO emissions from Tier 2 vehicles needs to be confirmed through analysis of data, including those from cold starts at 0°F and 10°F. Again, testing should include CO as well as other pollutants. If the analysis of Tier 2 and prior controls indicates that all locations will attain the 8-hour CO standard, more stringent federal CO vehicle emissions standards will be unnecessary. The results of all emissions testing must be incorporated into EPA's MOBILE model to accurately estimate future CO emissions. The effects on CO problem areas of the sales strategy used by manufacturers to meet the NO_x limits and of the trading and banking provisions also need to be assessed and incorporated into emissions modeling.

High-Emitting Vehicles

Findings

A relatively small number of high-emitting vehicles contribute disproportionately to CO and other motor-vehicle emissions. The vehicle fleets operating in and around places with high local concentrations of CO (hot spots) often include a higher proportion of high-emitting vehicles compared with the surrounding region. Elimination or repair of high emitters would

---

[8]The MOBILE model is used to estimate current and future on-road motor-vehicle emissions. MOBILE6 is the current version of that model.

likely reduce the severity of CO hot spots and reduce motor-vehicle emissions overall.

Recommendations

Air quality management agencies should identify high-emitting vehicles and target them for repair or removal from the fleet. Enhanced onboard diagnostic testing programs, tailpipe testing, motor-vehicle emissions profiling, and/or remote sensing can identify these vehicles. However, programs designed to mandate repair or removal of high-emitting vehicles might raise issues of fairness, because high emitters are often owned by people with limited economic means. The cost-effectiveness and equity impacts of policies that provide incentives for owners of high-emitting vehicles to seek repairs or vehicle replacement, such as repair assistance programs, should be explored. There should also be additional low-temperature testing to evaluate the effectiveness of programs aimed at controlling high-emitting vehicles. This evaluation should include the impacts on CO as well as other emissions.

Oxygenated Fuels

Findings

An oxygenated fuel is a gasoline containing an oxygenate (typically methyl *tertiary*-butyl ether [MTBE] or ethanol) intended to reduce production of CO. Oxygenated fuels program benefits are declining in effectiveness as more modern vehicles enter the fleet. EPA's MOBILE model predicts CO emissions reductions from oxygenated fuels of 3-7% for the 2010-2015 fleet because of reduced emissions from pre-1994 vehicles, cold starts, and malfunctioning vehicles. There is still uncertainty about the overall effectiveness of oxygenated fuels, especially at temperatures below 20°F.

Recommendations

EPA should undertake a science and policy review of the current oxygenated fuels programs to determine the conditions under which these programs are cost-effective. The review should also determine when changes in fleet technologies will render these programs ineffective. Low-temperature testing, especially below 20°F, is recommended. Oxygenated fuels programs should be implemented only when they provide cost-effective reductions in CO that help areas come into compliance or prevent areas that have attained the NAAQS from falling back into nonattainment.

Public Education

Findings

On the basis of its review of programs in Fairbanks, Alaska, the committee is concerned that public education campaigns have not sufficiently emphasized the adverse health effects associated with exposure to high ambient concentrations of CO. Also, the public is not fully aware of the link between transportation choices and overall air quality. As a result, public acceptance of and participation in locally proposed programs to achieve and maintain attainment of the NAAQS for CO is often poor.

Recommendations

Public-health education to improve public acceptance and compliance should be a component of all local emissions-reduction programs. Communities should use surveys and focus groups to regularly evaluate the effectiveness of public education programs and the impact they have on the success of CO emissions control.

CO Assessment

CO As an Indicator of Motor-Vehicle Pollutants

Findings

In urban environments, ambient CO concentrations are a strong indicator of motor-vehicle emissions. EPA estimates that as much as 95% of all CO emissions in some cities can be from automobile exhaust. Spatial and temporal variability in motor-vehicle activity and atmospheric dispersion characteristics can lead to CO hot spots.

CO is useful as a gauge of human exposure to other directly emitted mobile-source pollutants, such as air toxics and $PM_{2.5}$. However, CO is not a perfect indicator of all mobile-source pollutant emissions, because CO reacts more slowly than many other pollutants, and the ratio of CO to copollutants varies by emissions source. The atmospheric conditions that produce high CO concentrations are different from those that produce high concentrations of photochemical pollutants. Despite these caveats, measurements in the Los Angeles area and elsewhere have shown strong correlations between ambient CO and benzene concentrations. A strong correlation between CO and concentrations of the relatively short-lived 1,3-butadiene also was observed.

Recommendations

The committee has several recommendations in regard to the use of CO to represent the distribution of other pollutants. CO can be used to demonstrate the spatial distribution of some mobile-source pollutants, to identify hot spots, and to improve model representation of relationships between transportation activity and emissions. CO can also be used to approximate the concentrations of some air toxics arising from motor-vehicle exhaust emissions, such as benzene, 1,3-butadiene, and perhaps directly emitted $PM_{2.5}$. CO is most useful as an indicator in the microscale setting where concentrations of pollutants vary dramatically over short distances (e.g., with distance from a roadway). It is less reliable in representing regional distributions of these pollutants and is probably a poor indicator of motor-vehicle air toxics, such as formaldehyde and acetaldehyde, that react rapidly and have substantial sources in the atmosphere.

Spatial Distribution of CO

Findings

Although ambient CO concentrations have dropped considerably throughout the country, the number of monitors is inadequate to characterize CO distribution and identify all locations of high CO concentrations. There may be hot spots within cities that have already attained the NAAQS for CO or have not previously registered high ambient CO concentrations. The locations of these hot spots may raise social equity issues regarding exposure to mobile-source-related pollutants.

The monitors that have registered CO concentrations in excess of the NAAQS since 1995 are predominantly located in lower-income areas with greater minority populations. However, unidentified hot spots might exist in any location. Current data are insufficient to adequately characterize the relationships between hot spot locations, population characteristics, and health impacts.

Recommendations

EPA should employ air quality modeling and saturation studies[9] in CO problem areas to better characterize the spatial distribution of CO and the populations affected. The information garnered can be used to improve site selection for permanent monitoring, to improve model performance, and to address possible environmental equity issues. Programs targeted to local conditions can be developed using this information. These results should also be linked to health impact studies in these locations. In particular, EPA should try to better understand the upper end (higher CO levels) of the distribution of ambient exposures to motor-vehicle emissions that occur in most CO hot spots.

Permanent Monitoring

Findings

Although fewer and fewer locations are experiencing CO concentrations above or near the NAAQS, the continued operation of most current

---

[9]Saturation studies typically rely on portable monitors that "saturate" a geographical area with samplers to assess the air quality in places where high concentrations of pollutants are possible. Monitors can be deployed at temporary fixed-site locations or in mobile sampling vehicles. These studies are helpful to air pollution control agencies for evaluating their ambient air monitoring networks, characterizing pollutant concentrations over the entire saturation study area, and locating hot spots or high pollutant impact points. Personal monitoring could be incorporated into such studies to relate ambient concentrations to personal exposure.

ambient monitors remains essential for long-term assessments of air quality and health impacts.

Recommendations

Because of the value of CO monitoring information for air quality management in general, agencies should resist removing CO monitors in locations not expected to show violations. Instead, they should consider continuing operations at existing CO monitoring sites, noting that when monitors are co-located the incremental costs of continued operation may be relatively small compared with the data's usefulness for purposes beyond demonstrating attainment. The number and placement of permanent monitors also need to reflect changes in growth and development patterns to accurately assess the local air pollution situation. However, communities that have attained the CO standard with an adequate level of protection of safety might not be willing to pay to obtain data from these monitors. Support from federal and other sources might be needed to continue monitoring operations.

Ambient CO Modeling

Findings

Emissions and air quality models are important tools for air quality planning. Models help forecast changes in the mass of pollutants emitted resulting from controls and severe air pollution events. Models are also used to demonstrate attainment of the CO NAAQS, evaluate the effects of new construction projects that greatly increase emissions, and research the causes of pollution episodes and how to predict them. However, the spatial and temporal resolution of models typically used in CO management at this time is too coarse to capture the variability in pollutant concentrations, which is necessary to identify local hot spots and accurately represent unusual meteorological conditions.

Statistical forecasting models have been used to assess the probability of future high-CO episodes. The approach was used in Denver, Colorado, to assess the probability of having CO concentrations in excess of the NAAQS after the alteration of the oxygenated fuels program. This model takes into account the historical variability in CO concentrations resulting from meteorology and unusual traffic events.

Recommendations

More sophisticated, physically comprehensive models that can simulate how CO concentrations vary in time and space should be developed, applied, and evaluated. Ongoing research should be continued. Such models would be used for air quality planning and forecasting and for assessing human exposure to high concentrations of CO and related pollutants. Because CO is a relatively unreactive pollutant, the ability to better represent CO's temporal and spatial distribution provides an effective diagnosis of atmospheric dispersion patterns. Model improvements would have applications for other air quality management issues and would offer the potential to better understand the dispersion of chemical, biological, and radiological materials. Most important, improved models will permit more effective and realistic planning, leading to better-informed decisions by administrators. Model development should occur in concert with improved monitoring to enable model evaluation. In addition, the statistical forecasting models should be improved.

1

Ambient Carbon Monoxide Pollution in the United States

INTRODUCTION

Carbon monoxide (CO) has been central to the evolution of air quality management in the United States. CO is produced primarily by the incomplete combustion of carbon-containing fuels, such as gasoline, natural gas, oil, coal, and wood. In a 1977 National Research Council (NRC) report, CO was declared "probably the most publicized and best known criteria pollutant" (NRC 1977). The NRC attributed this recognition to the severe adverse health effects (including death) that result from acute exposure, which have been observed for centuries. Reducing human exposure to the products of incomplete combustion was an early objective of air quality management in the United States. CO was and still is the most recognizable indicator of incomplete combustion and has long been viewed as one of the most fundamental indicators of ambient air quality. When continuous monitors were first installed in some cities in the early 1960s, maximum 8-hour average concentrations in excess of 30 parts per million (ppm) were not unusual (DHEW 1970, EPA 1979).

National Ambient Air Quality Standards (NAAQS) for ambient concentrations of CO (9 ppm for an 8-hour average and 35 ppm for a 1-hour average) were instituted in 1971 on the basis of studies linking ambient CO concentrations with neurobehavioral effects. Although neurobehavioral effects no longer serve as the basis for the standards, subsequent studies linking CO to increased risk of chest pain and hospitalization for persons with coronary artery disease have supported retention of the regulation.

In the early 1970s, monitoring of CO indicated that exceedances of the 8-hour standard were common. The first comprehensive national report on emissions and air quality trends found that over 90% of monitors operating in 1971 recorded exceedances of the 8-hour NAAQS (EPA 1973). However, the situation improved fairly rapidly, primarily due to vehicle pollution controls. By the year 2000, only four locations (Birmingham, Alabama; Calexico, California; Lynwood, California; and Fairbanks, Alaska) reported exceedances of the 8-hour standard (Table 1-1). As of the end of 2002, both Lynwood and Fairbanks reported 2 years with no violation of the CO standard.

The locations that continue to have high concentrations of CO also tend to have topographical and meteorological characteristics that exacerbate pollution (e.g., nearby hills that inhibit wind flow and temperature inversions that inhibit vertical pollutant dispersion). Attainment of the health-based NAAQS for CO has proved somewhat difficult under those conditions. The question arises whether unique approaches are necessary to manage CO in such problem areas or the current policies will ultimately achieve good air quality. An issue for areas that now meet the NAAQS is their vulnerability to future exceedances as a result of increases in vehicle-miles traveled (VMT) or unusual meteorological conditions favoring CO accumulation.

STUDY BACKGROUND AND CHARGE

In response to the challenges posed for some locations by the NAAQS for CO, a committee was established by the NRC to investigate the problem of CO in areas with meteorological and topographical problems. The committee's statement of task was as follows:

An NRC committee will assess various potential approaches to predicting, assessing, and managing episodes of high concentrations of CO in meteorological or topographical problem areas. The committee will consider interrelationships among emission sources, patterns of peak ambient CO concentrations, and various CO emissions control measures in such areas. In addition, the committee will consider ways to better understand

TABLE 1-1 Data[a] for the Cities with the Greatest Numbers of CO Exceedance Days, 1995-2001

City and State	Monitor AIRS ID[b,c]	Number of Days with Exceedances of 8-hour CO Standard							Total 1995-2001	Latitude[d]	Longitude[d]	Elevation Above MSL (ft)
		1995	1996	1997	1998	1999	2000	2001				
Birmingham, AL - Shuttlesworth and 41st	01-073-6004	NA	2	2	6	17	9	33	69	33 33 54N	86 47 48W	580
Calexico, CA - 129 Ethel St.	06-025-0005	15	9	10	6	11	6	2	59	32 40 33N	115 28 59W	3
Lynwood, CA - Long Beach Blvd.	06-037-1301	12	17	10	9	8	2	0	58	33 55 45N	118 12 35W	88
Fairbanks, AK - Cushman	02-090-0002	8	1	2	2	2	1	0	16	64 50 43N	147 43 16W	460
Phoenix, AZ - Grand Ave. and Thomas Rd.	04-013-0022	3	2	1	0	1	0	0	7	33 28 45N	112 06 45W	NA
Spokane, WA - Third Ave.	53-063-0044	4	1	NA	NA	NA	NA	NA	5	47 39 13N	117 25 07W	1,970
Spokane, WA - Hamilton St.	53-063-0040	0	1	0	0	0	0	0	1	47 40 10N	117 23 43W	1,900
Las Vegas, NV - East Charleston Blvd.	32-003-0557	1	3	0	NA	NA	NA	NA	4	36 09 32N	115 06 36W	1,860

City - Monitor	AIRS ID									Latitude	Longitude	Elevation (MSL)
Las Vegas, NV - Sunrise Ave.	32-003-0561	NA	0	1	2	0	0	0	3	36 09 47N	115 06 52W	NA
Anchorage, AK - 3201 New Seward Hwy	02-020-0037	0	3	0	0	0	0	0	3	61 11 38N	149 52 00W	1,332
Anchorage, AK - 3201 Turnagain	02-020-0048	NA	NA	NA	0	1	0	1	2	61 11 32N	149 59 09W	66
El Paso, TX - North Campbell	48-141-0027	0	2	1	0	0	0	0	3	31 45 46N	106 29 12W	3,740
Kalispell, MT - Idaho and Main	30-029-0045	0	2	0	0	0	0	0	2	48 12 08N	114 18 49W	2,956
Denver, CO - Broadway - Camp	08-031-0002	2	0	0	0	0	0	0	2	39 45 04N	104 59 14W	5,220

[a] Data were provided by Laurence Elmore and Jake Summers (EPA). Calexico exceedance days in 2001 were from Marcella Nystrom (California Air Resources Board), and are only for December.
[b] The numbers in the EPA's AIRS (aerometric information retrieval system) ID refer to the state, county, and individual monitor codes, respectively.
[c] Only monitors showing exceedances in each city are listed.
[d] The numbers for latitude and longitude are in degrees, minutes, and seconds of angle north of the equator and west of Greenwich.

Abbreviations: MSL, mean sea level; NA, not available.

relationships between episodes of high ambient CO, personal exposure, and the public-health impact of such episodes and alternative ways to measure progress in controlling ambient CO. An interim report dealing with Fairbanks, Alaska, as a case study was completed in May of 2002. A final report, including other CO problem areas, will be completed by the end of the study.

The committee will address the following specific issues:

- Types of emission sources and operating conditions that contribute most to episodes of high ambient CO.
- Scientific bases of current and potential additional approaches for developing and implementing plans to manage CO air quality, including the possibility of new catalyst technology, alternative fuels, cold-start technology, as well as traffic and other management programs for motor vehicle sources. Control of stationary source contributions to CO air quality also will be considered.
- Assessing the effectiveness of CO emissions control programs, including comparisons among areas with and without unusual topographical or meteorological conditions.
- Relationships between monitored episodes of high ambient CO concentrations and personal human exposure.
- The public-health impact of such episodes.
- Statistically robust alternative methods to assist in tracking progress in reducing CO that bear a relation to the CO concentrations considered harmful to human health.

This study provides scientific and technical information potentially helpful in the development of state implementation plans (SIPs); however, the committee does not provide prescriptive advice on the development of specific SIPs for achieving CO attainment. In addition, the committee does not suggest changes in regulatory compliance requirements for areas in nonattainment of the NAAQS, and it does not recommend changes in the NAAQS for CO.

Fairbanks, Alaska, was chosen as a case study for the interim report because its meteorological and topographical characteristics make it susceptible to severe winter inversions that trap CO and other pollutants at ground level (NRC 2002).

SUMMARY OF INTERIM REPORT

Fairbanks, Alaska, was chosen as a case study for the interim report because its meteorological and topographical characteristics make it susceptible to severe winter inversions that trap CO and other pollutants at ground level. The committee's interim report, entitled *The Ongoing Challenge of Managing Carbon Monoxide Pollution in Fairbanks, Alaska*, was completed in 2002 (NRC 2002). The interim report allowed the committee to assess the characteristics of the CO problem in detail in one meteorological and topographical problem area. It provided the committee with general lessons applicable to other locations as well as characteristics of the problem that were unique to Fairbanks. Because the committee devoted significant effort to assessing CO episodes in Fairbanks, the final report also draws on the Fairbanks case study.

In the interim report, the committee found that Fairbanks has made great progress in reducing its violations of the 8-hour CO health standard, and has reduced the number of days annually with violations from over 130 during 1973 and 1974 to zero over the last 2 years (2001 and 2002). Despite this progress, the committee also concluded that Fairbanks will continue to be susceptible to violating the 8-hour CO health standard on some occasions for many years to come because of its unfavorable meteorological and topographical conditions. Those adverse natural conditions might be compounded by future increases in the population of Fairbanks brought about by large pipeline or military construction projects. The findings associated with the progress on reducing CO violations and vulnerability to future violations are relevant to CO episodes in other meteorological and topographical problem areas.

The committee recommended that there be improvements to local emissions controls, including vehicle emissions inspection and maintenance (I/M) programs, low-sulfur gasoline, and traffic control strategies. In particular, the committee found that the Fairbanks North Star Borough is making substantial efforts to characterize and control cold-start emissions, despite the difficulty in quantifying emissions-reduction credits in its CO-attainment plan. The main method for controlling these emissions is through electrical heating devices known as plug-ins that preheat the engine coolant or lubricant of parked motor vehicles. Plug-ins substantially reduce CO emissions during the cold-start phase of engine operations by reducing the length of time needed for the catalyst to become fully operational. The committee recommended that the borough continue to expand the plug-in program by requiring or encouraging the equipping of more parking spaces with electric outlets for plug-ins. The committee noted that the alert-day program, where the public is alerted on days when CO is forecasted to exceed the standard, was important in Fairbanks because it was part of a larger public information campaign to encourage motorists to use plug-ins. The emphasis on controlling cold-start emissions through plug-ins is generally limited to Fairbanks and likely does not translate well to the lower 48 states.

In its interim report, the committee noted that officials with the borough argued that Fairbanks should be granted an exemption from the Clean Air Act with regard to the ambient CO health standards because of its extreme meteorological and topographical conditions. However, the committee concluded that a similar argument could be made for other regions with regard to a variety of air pollutants. Furthermore, the ambient concentrations of CO observed in Fairbanks have exceeded the level that EPA identified in the health-based standard for the protection of the general population and susceptible individuals. Thus, the committee concluded that efforts to control CO be continued and improved.

THE COMMITTEE'S APPROACH TO ITS CHARGE

In this final report, the committee addressed meteorological and topographical conditions that foster pollution episodes, CO and related emissions from mobile and other sources, and air quality management options. The committee also examined monitoring data for ambient CO, including episodes when 8-hour average concentrations exceeded the NAAQS. In general, although monitoring of CO at locations with problems has occurred for an extended period of time, data and modeling to assess the spatial and temporal extent of high-CO events are limited. The limited data reduced the committee's ability to assess human exposures during high-CO episodes. In addition, little exposure or epidemiological data specific to locations with high-CO episodes resulting from meteorological and topographical conditions were available to assess the public health effects of those episodes. In the absence of such data, the committee reviewed relevant clinical and epidemiological studies presented in the scientific literature and considered by EPA in assessing the health effects of exposure to ambient CO at concentrations and durations exceeding the NAAQS. Thus the committee did not conduct a comprehensive examination of the health effects of CO and did not examine the likelihood or actual risk of adverse health effects from exposure to ambient CO concentrations in these meteorological and topographical problem areas. In addition, the committee did not consider other CO sources such as cigarettes and therefore did not attempt to put the risks of ambient CO exposure into the context of other CO sources.

The technical feasibility and potential for emissions reductions of a number of air quality management options are also discussed in the body of this report. Promising options available for controls at the federal, state, and local levels are presented in the summary. The committee found that light-duty vehicles are the primary source of emissions and potential emissions reductions, so most options focused on control of emissions from those vehicles. The committee's recommendations follow the relevant supporting evidence.

REPORT CONTENTS

This final report documents the committee's response to the charge described above. The report consists of four chapters and a summary. Chapter 1 provides background information on the regulation and sources of ambient CO pollution in the United States necessary to characterize the key issues stated in the committee's charge. The main topics include the NAAQS and areas exceeding CO standards, sources of CO emissions, health effects of CO, relationship of CO to other air pollutants, and issues relating to the spatial distribution of CO. Chapter 2 describes the meteorological and topographical conditions that foster pollution episodes in CO problem areas. Temperature inversions, long-term meteorological trends, temporal patterns of CO concentrations, and vulnerability of areas to future exceedances are discussed in detail. Chapter 3 discusses CO management and the tools needed to implement CO standards, such as emissions control strategies and monitoring and modeling tools. Some control strategies described include emissions standards, I/M programs, fuels, and transportation-control measures. Finally, Chapter 4 focuses on long-term issues related to exposures, controls, and management of CO.

NATIONAL REGULATORY SETTING FOR AMBIENT CO

National Ambient Air Quality Standards

To control adverse health effects from CO exposure, the U.S. Environmental Protection Agency (EPA), acting per Sections 108 and 109 of the Clean Air Act (CAA), established the NAAQS for CO in 1971. Recognizing that exposure can have both acute and longer-term effects, the NAAQS for CO have two criteria with different averaging periods: 35 ppm averaged over 1 hour, and 9 ppm averaged over 8 hours.[1] Each criterion is not to be exceeded more than once per year; the second and subsequent exceedances within a year are considered violations of the standard. The 8-hour standard has proven to be more difficult to meet than the 1-hour, especially for a handful of cities. The standard has been periodically reviewed on the basis of new scientific findings, as mandated by the CAA. The most recent review was published in 2000 (EPA 2000a).

EPA originally designated an area as being in "nonattainment" of the 8-hour standard if the second-highest 8-hour average CO concentration measured during a calendar year (known as the "design value") was greater than 9 ppm. After the Clean Air Act Amendments of 1990 (CAAA90), EPA designated areas that had previously been in nonattainment as "serious" if the design value was 16.5 ppm or greater, "moderate" if the design value was 9.1-16.4 ppm, and "not classified" if recent data were insufficient to determine whether the standard was met. Moderate areas that did not reach attainment by July 1996 could be reclassified by EPA as serious. Nonattainment areas are required to submit a state implementation plan (SIP) to EPA that includes a characterization of pollutant concentrations and emissions, a description of the emissions reductions the area plans to make, and an "attainment demonstration" showing how the emissions reductions will enable the area to attain and maintain compliance with the NAAQS. To be eligible for reclassification from nonattainment to attainment status for CO, an area must have air quality monitoring data indicating that it did not violate the NAAQS during the previous 2 years. Though areas typically apply for reclassification immediately, Smith and Woodruff (2001) discussed how Fort Collins, Colorado, has delayed their application for reclassification since 1994 to pursue wider air quality goals. To meet the city's goal of continually improving air quality as described in the City of Fort Collins Air Quality Action Plan (2001), the city used their nonattainment status to pursue emissions control strategies that are not traditionally implemented in attainment areas. One example is the vehicle emissions I/M program, which currently is not required when a city is in attain-

---

[1] Though the standard is 9.0 ppm, in practice an exceedance does not occur until the 8-hour average is greater than 9.4 ppm. Values between 9.1 ppm and 9.4 ppm are rounded down to 9.0 ppm.

ment. However, Fort Collins has recently applied to be reclassified and may discontinue some emissions control programs. Officials in Fort Collins are undertaking a feasibility study to explore both the voluntary and/or mandatory control programs after the state mandatory I/M program is eliminated.

Throughout the report, the terms exceedance, exceedance days, and violation are used. They are defined as follows: an exceedance of the CO standard is any CO concentration measurement of 9.5 ppm or above for an 8-hour average;[2] exceedance days are days on which one or more nonoverlapping 8-hour average CO concentration was 9.5 ppm or greater; and a violation is two or more exceedances within a calendar year. Note that more than one exceedance can occur in a day if there is more than one nonoverlapping 8-hour period with an average CO concentration of 9.5 ppm or greater.

Vulnerable Areas and the Form of the CO Standard

The form of the CO standard, where a violation occurs upon the second and all subsequent exceedances in a calendar year, contributes to the difficulties that meteorological and topographical problem areas have in attaining the standard. A significant probability of an exceedance exists with the current attainment test because of the stochastic nature of ambient air pollutant concentrations (Gibbons 2002). Areas must control CO under very infrequent, though not uncommon, meteorological conditions. The form of the new 8-hour ozone standard (the 3-year average of the fourth highest annual value) was changed from the form of the 1-hour standard (the fourth highest value over 3 years) in part because the latter form is more susceptible to extreme meteorological conditions.

Conformity Requirements

Transportation conformity requirements were originally developed to ensure that federal funding and approval was given to those transportation

---

[2]Although the 8-hour NAAQS for CO is 9 ppm, because early monitoring instruments had limited precision of about 1 ppm it has been the practice to consider an 8-hour average an exceedance only if it is 9.5 ppm or greater.

activities consistent with air quality goals. Transportation conformity was first introduced in the CAA Amendments of 1977, which included a provision linking air quality to transportation planning by ensuring that transportation investments conform with SIPs (DOT 2000). Conformity requirements were made more rigorous in the CAAA90 and in the regulations EPA issued in 1993 to implement the requirements (40 CFR § 51 and 93 [1993]). The CAAA90's conformity mandate requires that transportation plans, programs, and projects in nonattainment or maintenance areas funded or approved by the Federal Highway Administration (FHWA) or the Federal Transit Agency (FTA) do not: (1) create new violations of the federal air quality standards; (2) increase the frequency or severity of existing violations; or (3) delay timely attainment of CAA standards.

Metropolitan planning organizations (MPOs) are responsible for performing air quality conformity analyses. MPOs must have transportation plans in place that present a 20-year perspective on transportation investments for their region as well as a short-term transportation improvement program (TIP). The TIP is a multi-year prioritized list of projects (3 years at a minimum) proposed to be funded or approved by FHWA or FTA. The conformity analysis is done for the system of projects contained in a region's TIP and transportation plan, and must show emissions consistent with those allowed in the SIP.

Conformity determinations must be made at least every 3 years, or as changes are made to plans, TIPs, or projects. A formal interagency process is required to establish procedures for consultation between MPOs, EPA, FHWA, FTA, and state and local transportation and air quality agencies. These procedures apply to the development of the SIP, the transportation plan, the TIP, and conformity determinations. The SIP must establish interagency consultation procedures for the coordinating agencies and include schedules for implementation of all strategies. Once EPA approves the part of the SIP that describes the interagency consultation process (the conformity SIP), it is then enforceable by EPA as a federal regulation.

One of the key components of the conformity determination for CO is the application of project-level emissions analysis and, on occasion, hot-spot analysis. During SIP preparation, emissions budgets are created for nonattainment areas. These budgets set limits on the mass of the criteria pollutant that can be emitted in the area and are usually broken into general and transportation budgets. For an area to be in conformity with the SIP, the sum of the emissions from all transportation projects may not exceed the transportation budget., unless reductions in the general budget are made to compensate In cases where a project could create a local violation or exacerbate pollution in an existing problem area, hot-spot analysis also might be needed.

In hot-spot analysis, microscale CO concentrations resulting from an individual roadway project are modeled to investigate whether the project will cause a localized CO problem. Analysis of CO hot spots can be done quantitatively, typically through the use of Gaussian dispersion models, which are described in Chapter 3 of this report. Dispersion modeling may be needed to identify possible violations during SIP preparation. Also, traffic-simulation models can be combined with instantaneous emissions at the microscale level to predict emissions inventories and to assess queuing and traffic flow along specific roadway segments or at specific intersections. Proposed projects to change traffic patterns are often analyzed by starting with the three intersections with the highest traffic volumes and poorest level-of-service to determine if CO problems exist, and then modeling other intersections where capacity is equaled or exceeded. With EPA approval, areas also can establish their own thresholds for quantitative analysis (Guensler et al. 1998).

Alternative screening methods can be used for CO project-level hotspot analysis (40 CFR § 51 and 93 [1993]). Screening tools are simple estimation techniques that determine whether transportation projects are in need of more rigorous testing and additional analysis. They are used to provide conservative estimates of the air quality impacts of a specific source, with the assumption that if a project passes the conformity criteria using the screening tools then it would also pass more rigorous analysis. The benefit of screening tools is that they reduce the number of transportation projects requiring more detailed quantitative CO modeling and eliminate the need for more detailed modeling for those sources that clearly will not cause or contribute to an exceedance of the CO NAAQS.

AREAS WITH RECENT EXCEEDANCES OF THE CO STANDARD

Nationally, CO concentrations have declined significantly over the past 30 years. In the early 1970s, when CO monitoring in the United States became widespread, many cities reported numerous exceedances of both the 1-hour (35 ppm) and 8-hour (9 ppm) NAAQS for CO. In 1971, 53 of 58 monitoring stations (91%) recorded exceedances of the 8-hour standard, and 7 of 58 stations (12%) recorded exceedances of the 1-hour standard (EPA 1973). Improvements occurred rapidly, primarily resulting from advances in motor-vehicle emissions control technology. EPA (1976) noted that, although ambient concentrations of many other pollutants showed few signs of improvement, "there was an evident decline in the proportion of stations at which the 8-hour CO standard was exceeded." In 1974, the number of stations reporting exceedances of the 8-hour standard fell to 56% (211 of 377 stations). Since then, the number of areas showing exceedances of the 8-hour standard has continued to decrease, and no monitor has shown an exeedance of the 1-hour standard since 1995. EPA (2002a) reports that the national average ambient CO concentration in 2001 was 62% lower than it was in 1982 and 38% lower than it was in 1992.

Table 1-1 shows the 11 cities that have had the most difficulty meeting the 8-hour NAAQS since 1995, ranked by the total number of exceedance days at the monitor recording the highest number of exceedances during the 7-year period from 1995 to 2001. The table shows the aerometric information retrieval system identification number (AIRS ID), latitude, longitude, and elevation above mean sea level of each monitor. Second monitors are listed for Spokane, Washington, and Las Vegas, Nevada, because the first monitor ceased operating before the end of the 7-year period. A second monitor is listed for Anchorage, Alaska, because the second monitor is located in a residential neighborhood, rather than in a downtown area, providing an interesting diurnal comparison, and because it recorded the most recent exceedance in Anchorage.

Birmingham, Alabama, stands out in Table 1-1 because of the large total number of exceedance days during the 1995-2001 period and because the annual number appears to be increasing with time. This is a special case involving CO emissions from an industrial source—a mineral-wool facility. This industry is not regulated for CO emissions; however, a monitor placed close to the facility by the Jefferson County Public Health Department detected frequent exceedances of the CO standard. To rectify the problem, the facility has agreed to changes in their stack height and operating procedures. The committee did not become aware of the Birmingham exceedances until after its last meeting. The committee relied on EPA's listing of locations that have recently violated the CO standard in their annually updated air quality update (EPA 2002b). Birmingham, Alabama, was not listed as violating the CO standard in the air quality update until 2002, despite the fact that violations of the CO standard dated back at least 4 years. Because this issue came up so late in the study and because the exceedances are due to problems in the facility's operations, not meteorological and topographical features, Birmingham will not be discussed in detail in this report. However, this case indicates that other areas may be experiencing exceedances of the CO standard that are not being detected by the fixed monitoring site network.

Trends in national average ambient CO concentrations do not always mirror trends in nationwide emissions. One reason might be that most monitors are located in urban areas, so changes in air quality are most likely to track changes in urban air emissions rather than in total emissions. Because light-duty vehicles dominate urban emissions and air quality monitor sites are located near roadways, the improvements in ambient CO concentrations disproportionately reflect reductions in emissions from these vehicles, while emissions from most other sources remain basically unchanged.

Characteristics of Exceedances

Thirty years ago exceedances of the 8-hour NAAQS for CO occurred in all months of the year, but now they are a winter phenomenon in most areas.[3] Figures 1-1 and 1-2 show the number of days with exceedances by month and year (the year is defined as July through June centering on the winter season) for Lynwood, California, and Fairbanks, Alaska. There are two reasons for the pattern: reduced solar heating during winter, which favors a more stable atmosphere with less vertical mixing and lower windspeeds; and increased emissions in winter.

CO Trends in Problem Areas

Table 1-1 shows a generally downward trend in the number of exceedance days recorded in the 10 cities that are the focus of this report for the years since 1995. A downward trend in concentrations also can be seen in Figure 1-3, which shows the decline in the nationwide composite average of annual second-highest 8-hour CO concentrations from 1978 to 1997. The smoother decline of the composite average is a result of the large number of sites included.

The annual second-highest nonoverlapping 8-hour average CO concentration is a statistic that shows a great deal of variability at any one site. In its interim report (NRC 2002), the committee looked at trends in other

---

[3]Exceedance days occurred in Birmingham in June and July of 2001.

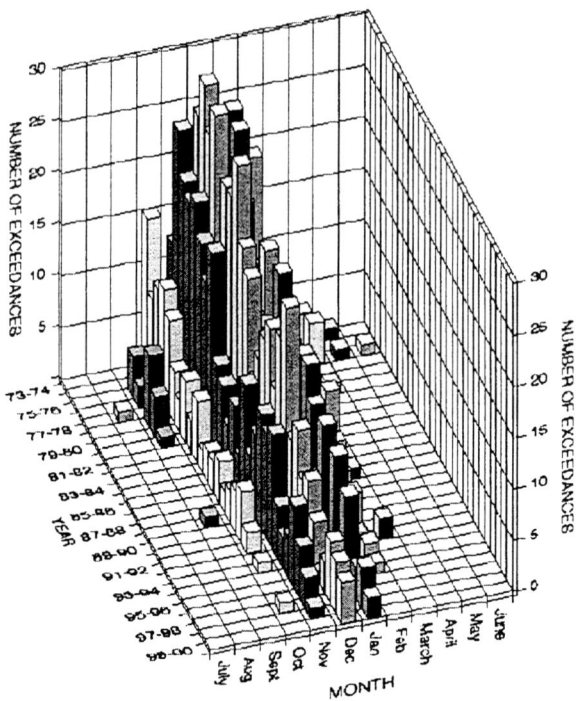

FIGURE 1-1 Number of days with exceedances of the 8-hour NAAQS for CO per month and per year in Lynwood, California.

statistics (including the 75th, 90th, 95th, and 99th percentiles) for the three monitoring sites in Fairbanks, Alaska, for the six winter seasons from 1995-1996 to 2000-2001. Although these other statistics showed less variability, they could all be fit over the 6-year period with straight lines. The slopes of the lines showed that the statistics declined at about 7% per year, consistent with steadily declining CO concentrations. A recent study (Eisinger et al. 2002) looked at whether the downward trend in CO concentrations was also occurring at microscale monitoring sites—sites located in close proximity to high traffic density. The study concluded that, although CO concentrations at the microscale sites are often higher than concentrations found at larger-scale monitoring sites (sites located in extended urban areas and more rural areas), CO concentrations at microscale sites are declining at the same rate as concentrations recorded at monitors representing larger regions.

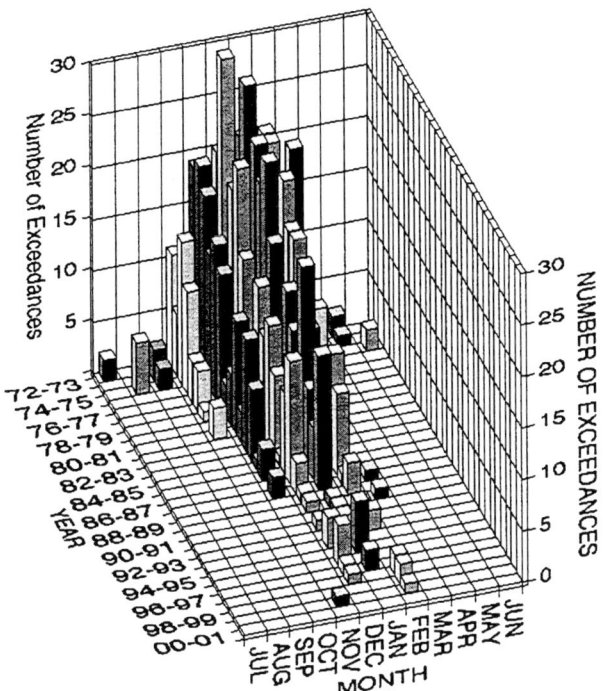

FIGURE 1-2 Number of days with exceedances of the 8-hour NAAQS for CO per month and per year in Fairbanks, Alaska. The numbers include days on which exceedances occurred at any of the three monitoring sites.

SOURCES OF CO EMISSIONS

National Inventory

The major categories of CO emissions sources include transportation (mobile sources), industrial processes, nontransportation fuel combustion (which includes stationary and area sources), and miscellaneous sources. Figure 1-4 presents an estimate of national CO emissions over the 1982-2001 time period. The figure gives a general indication of the dominant fraction of mobile-source emissions compared with other major source categories.

The mobile-source emissions referred to in Figure 1-4 can be separated into on-road, off-road, and nonroad emissions. On-road emissions come from both light-duty vehicles (LDVs) and heavy-duty vehicles (HDVs).

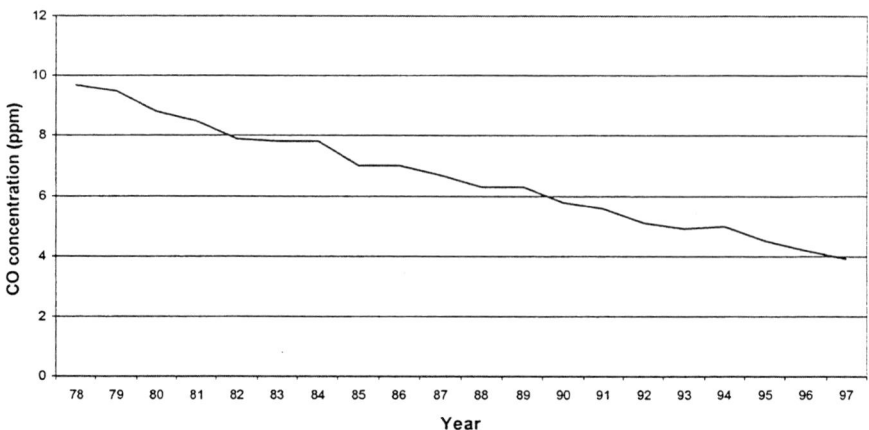

FIGURE 1-3 Nationwide composite average of the annual second-highest 8-hour CO concentrations, 1978-1997. Note that there were 184 monitoring sites from 1978 to 1987 and 368 sites from 1988 to 1997.

The boundary separating LDVs from HDVs historically has been 8,500 pounds gross vehicle weight (GVW) (the weight of the vehicle plus the weight of the rated load-hauling capacity). LDVs, which include passenger cars and light trucks, are fueled primarily by gasoline; HDVs use both diesel fuel and gasoline. The heavier HDVs (those with a GVW greater than 26,000 pounds) use diesel fuel almost exclusively. Diesel engines emit much less CO overall compared with LDVs; thus, on-road CO inventories are dominated by LDV emissions. Nonroad emissions come from nonroad engines including construction, logging, mining, and farm equipment and lawn and garden equipment. Off-road vehicles include marine vessels, recreational vehicles, locomotives, and aircraft.

Table 1-2 provides an estimated inventory of CO emissions in the United States in 1999 (EPA 2001a). An estimated 77% of the anthropogenic CO emissions come from mobile sources, including on-road vehicles (51%) and nonroad engines and vehicles (26%). The remaining CO emissions are from area and point sources, including fuel combustion and industrial processes. It should be noted, however, that this inventory has significant uncertainties. For example, an NRC report (2000) reviewing EPA's Mobile Source Emissions Factor (MOBILE) model discusses the

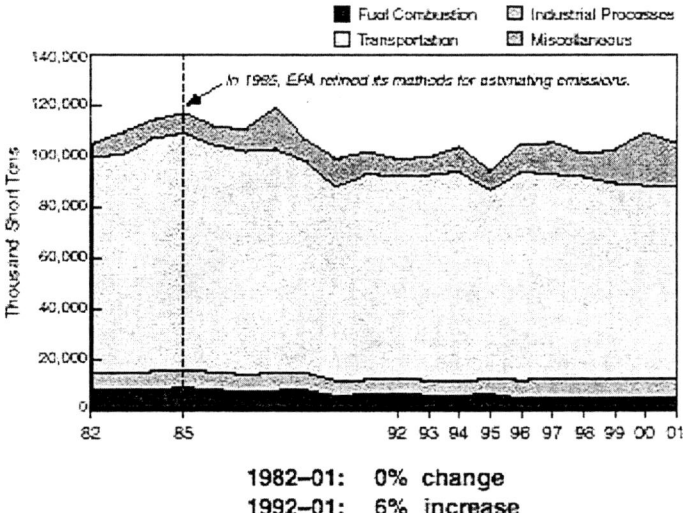

FIGURE 1-4 Nationwide CO emissions from 1982 to 2001. Note that from 1982 to 2001 there was 0% change, and from 1992 to 2001 there was a 6% increase in emissions. Emissions are shown in thousands of short tons. One short ton is equivalent to 2,000 pounds (lb) or 0.9072 metric tons. A long ton is a measurement weight equivalent to 2,240 lb or 1.0 metric tons. Source: EPA 2002a.

substantial inaccuracies in estimates of fleet emissions and effectiveness of control strategies for on-road vehicles.

Regional Inventories

In urban areas, mobile sources tend to contribute more to the mix of emissions than indicated by the national average. On the basis of its MOBILE model, EPA suggests that vehicles may contribute 95% or more of CO emissions in cities classified by EPA as having serious air pollution (EPA 2001a). Table 1-3 shows emissions inventories for five cities that have had CO exceedances in the past. Mobile sources contribute most of the CO emissions, ranging from 78% for Fairbanks to 96% for Phoenix. On-road vehicles contribute the majority of mobile-source emissions, ranging from 62% to 84%.

TABLE 1-2 National CO Emissions Inventory Estimates for 1999

Source Category	Thousands of Short Tons	Percent of Total (%)
Point or area fuel combustion	5,322	5.46
Electric utilities	445	0.457
Industry	1,178	1.21
Residential wood burning	3,300	3.39
Other	399	0.409
Industrial processes	7,590	7.79
Chemical and allied product manufacturing	1,081	1.11
Metals processing	1,678	1.72
Petroleum and related industries	366	0.376
Waste disposal and recycling	3,792	3.89
Other industrial processes	599	0.615
On-road vehicles	49,989	51.3
Light-duty gas vehicles and motorcycles	27,382	28.1
Light-duty gas trucks	16,115	16.5
Heavy-duty gas vehicles	4,262	4.37
Diesels	2,230	2.29
Nonroad engines and vehicles	25,162	25.8
Recreational	3,616	3.71
Lawn and garden	11,116	11.4
Aircraft	1,002	1.03
Light commercial	4,259	4.37
Other	5,169	5.30
Miscellaneous	9,378	9.62
Slash or prescribed burning	6,152	6.31
Forest wildfires	2,638	2.71
Other	588	0.615
Total	97,441	

Source: EPA 2001a.

TABLE 1-3 Regional CO Emissions Inventories for Selected Cities (Tons Per Day)

Source Category	LA-South Coast, CA 2001	Percent	Fairbanks, AK 2001	Percent	Las Vegas, NV 1990	Percent	Anchorage, AK 1995	Percent	Denver, CO 2001	Percent
On-road vehicles	4,244	75.9	14.4	61.8	310	80.1	113	84.0	875.2	72.8
Nonroad engines and vehicles	882	15.8	3.7	15.7	60	15.6	12	9.0	161.5	13.4
Recreational	249				6					
Lawn and garden					4					
Aircraft	56		3.3		40				23.9	
Light commercial	5								136.6	
Other	572		0.4		11				1	
Area sources	323	5.8	0.9	3.8	10	2.6	5	4.0	96.6	8.0
Fuel combustion			0.1							
Industrial processes									25.1	
Other			0.8						71.5	
Point sources	53	0.9	4.3	18.6	7	1.7	1	1.0	70.2	5.8
Natural sources	89	1.6	0.0	0.0	0	0.0	0	0.0	0	0.0
Miscellaneous	0		0.0	0.0	0	0.0	4	3.0	0	0.0
Total	5,590	100.0	23.3	100.0	387	100.0	135	100.0	1,203.5	100.0

Regional inventories are important because they form the basis for SIPs and are used in assessing local projects. However, the use of regional emissions inventories for analysis of localized CO exceedances can be problematic because these inventories might include sources that do not contribute to exceedances at specific locations. Air intakes at CO monitoring sites are a few meters above street level, so sources at higher elevations (e.g., smokestacks) might emit CO at an elevation higher than the inversion level and contribute little to measured CO concentrations. For example, in its interim report (NRC 2002) the committee noted the presence of very strong ground-level inversions in Fairbanks, Alaska. Such conditions mean that local power-plant emissions released well above the inversion height likely do not mix with ambient air at the monitor height and, thus, do not contribute to the high CO concentrations recorded at monitoring sites. In addition, some CO sources included within these regional inventories might be located at great distances from monitors and might not contribute to local exceedances. These issues are discussed further in Chapter 4.

Vehicle Sources of CO Emissions

The primary cause of CO emissions from vehicles is the incomplete combustion of gasoline. The fuel-oxidation process (combustion) is the conversion of the fuel to lower-molecular-weight intermediate hydrocarbons (including olefins and aromatics). These hydrocarbons are converted to aldehydes and ketones, then to CO, and finally to carbon dioxide (CO_2). The initial reactions are faster than the final conversion from CO to CO_2. Incomplete conversion of fuel carbon to CO_2 results in part from insufficient oxygen in the combustion mixture—known as fuel-rich[4] conditions—and insufficient time to oxidize fuel carbon to CO_2. CO emissions from diesel-powered vehicles are minimal compared with emissions from gasoline-powered vehicles, primarily because excess air is used in the diesel combustion cycle. The excess air, combined with high temperatures and pressures involved in the diesel cycle ensures more complete combustion. Hence, the following discussion is limited to gasoline-powered vehicles.

---

[4]The ratio of air to fuel mass that provides just enough O_2 to convert all the carbon and hydrogen in gasoline to CO_2 and water (termed the stoichiometric ratio) is about 14.7 to 1. Ratios less than 14.7 to 1 have more fuel than is optimal and are called fuel-rich.

Trends in Vehicle Emissions

Vehicles produce excessive CO when cold starts, increased load (e.g., climbing a hill), rapid acceleration, or engine malfunctions induce fuel-rich conditions (NRC 2000). CO and unburned fuel can be greatly reduced by a variety of techniques, including additional oxidation in a catalytic converter mounted between the engine and the muffler. Box 1-1 lists some of the major milestones in the control of emissions from automobiles starting with the Clean Air Act (CAA) of 1970 (EPA 2001a). These milestones, which have led to large reductions in CO and other pollutant emissions, include national standards for tailpipe emissions, new vehicle technologies, and clean fuels programs as well as state and local vehicle emissions I/M programs and transportation management programs.

According to EPA, overall CO emissions were reduced by 25% between 1970 and 1998 (EPA 2000b). Light-duty gasoline-vehicle emissions have shown a 57% decrease. Per-vehicle emissions have been reduced even more. Substantial reductions in light-duty gasoline-vehicle emissions over the past 30 years have offset increases in CO emissions from other sources; however, light-duty gasoline-powered vehicles continue to dominate CO emissions inventories.

In addition to per-vehicle emissions, other factors—such as vehicle-miles traveled (VMT), population growth, trip making, and the rapid growth of sport utility vehicles (SUVs) in the vehicle population—impact total CO emissions. The improvement in air quality discussed earlier in this chapter occurred despite an approximate 35% increase in VMT in the United States during the 10-year period from 1992 to 2001 (EPA 2001b). Figure 1-5 illustrates the increases in VMT and the number of licensed drivers in the United States over the past 20 years and the decline in CO emissions from LDVs over the same period. Eventually, continuing increases in VMT might result in an increase in net CO emissions unless new emissions control technologies continue to reduce CO emission rates from new or in-use vehicles. Future CO emissions also might be affected by changes in the vehicle subfleet population as higher-emitting Tier 1 vehicles, such as SUVs, begin to make up a greater fraction of the overall fleet.

Other Sources of Ambient CO Emissions

The remaining CO emissions come from point and area sources, in-

> **BOX 1-1** Milestones in Motor-Vehicle Emissions Control
>
> 1968 Federal Air Quality Control Act allows California to enforce its own emissions standards for new vehicles.
> 1970 Clear Air Act (CAA) sets automobile emissions standards to be met in 1975.
> 1971 Charcoal canisters appear to meet evaporative standards (do not control CO).
> 1973 Exhaust gas recirculation valves improve nitrogen oxides (NO_x) control (do not improve CO).
> 1974 Fuel-economy standards are set.
> 1975 The first catalytic converters appear for hydrocarbons (HCs) and CO. Unleaded gas required for use in catalyst-equipped cars.
> 1977 Amendments to the CAA define and tighten vehicle standards and clarify the definition of nonattainment areas.
> 1981 Three-way catalysts with onboard computers and 0_2 sensors appear for HC, CO, and NO_x control.
> 1983 Inspection and maintenance (I/M) programs are established in 64 cities.
> 1989 Fuel volatility limits are set (do not control CO).
> 1990 CAA Amendments set new tailpipe standards beginning in 1994.
> 1990 California Air Resources Board approves low and zero emissions vehicles standards for California.
> 1992 Oxygenated fuels introduced in cities with high levels of CO.
> 1993 Limits set on sulfur content of diesel fuel.
> 1994 Phase-in begins of new Tier 1 vehicle standards and technology.
> 1994 Phase-in of cold-temperature (20°F) CO standard.
> 1995 Second generation onboard diagnostic systems (OBDII) required in 1996 model year cars.
> 1998 National low emissions vehicle program begins in the northeast with sales of 1999 model year California emissions equipped vehicles.
> 2000 Phase in of supplemental federal test procedure as vehicle certification test.
> 2004 Tier 2 standards begin.
>
> Sources: CARB 2003; EPA 2001a.

cluding wood burning, lawn and garden equipment, and natural sources, such as wildfires. It should be noted that CO is produced not only through incomplete combustion but also from the thermal decomposition of CO_2,

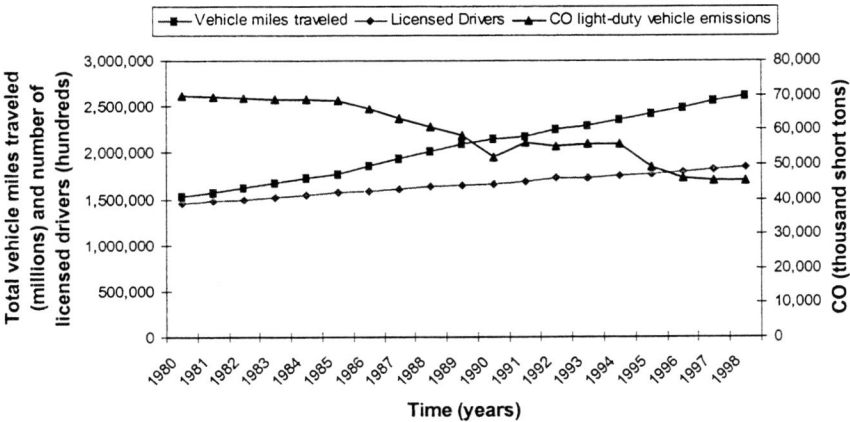

FIGURE 1-5 Nationwide trends in total vehicle-miles traveled, number of licensed drivers, and CO light-duty vehicle emissions from 1980 to 1998.

a process that occurs mainly in power plants. The contribution of CO produced through thermal decomposition to emissions inventories however is likely to be small. Table 1-2 shows nationwide CO emissions in 1999 from various sources. Area and point sources make up a smaller fraction of CO emissions inventories than mobile sources do, but are still important contributors to total CO emissions. For example, point and area sources can play a dominant role in localized exceedances, such as those that have occurred recently in Birmingham, Alabama. In addition, operators of lawn and garden equipment may experience high personal exposures to CO.

HEALTH EFFECTS OF CO

Clinical and Epidemiological Studies of CO Effects

The health effects of CO have been assessed through controlled exposure of human volunteers and a growing body of community epidemiological studies (EPA 2000a). Exposure studies executed under various experimental protocols have produced substantial information on the toxicity of CO, its direct effects on blood and tissues, and how those effects are expressed in terms of changes in organ functions. Community epidemiological studies attempt to extend these results to understand the potential community health effects of ambient CO exposures.

CO affects human health by impairing the ability of the blood to carry oxygen (O_2) to body tissues. When CO is inhaled, it rapidly crosses from the lungs into the blood, where it binds to hemoglobin to form carboxyhemoglobin (COHb), a useful marker for predicting CO health effects. Because CO has an affinity for hemoglobin more than 200 times that of O_2, the presence of CO in the lungs will displace O_2 from the hemoglobin. In other words, when CO is present in the lungs, hemoglobin is unable to reach complete O_2 saturation. In addition, the binding of CO increases hemoglobin's binding of O_2, thereby inhibiting the release of O_2 from hemoglobin to body tissues. The effect of COHb is illustrated by a leftward shift in the O_2-hemoglobin dissociation curve (Figure 1-6). This effect continues until the COHb dissociates, typically several hours after CO exposure. CO also may affect O_2 transport to muscle (EPA 2000a). CO has been shown to bind to myoglobin, which supplies oxygen to muscles during strenuous exercise, when muscle demand for oxygen is greater than the supply of oxygen available from the blood.

COHb levels in healthy individuals not exposed to high concentrations of ambient CO are typically 0.3% to 0.7%. CO is formed endogenously through normal metabolism of heme leading to approximately these levels of COHb (i.e., this occurs irrespective of ambient exposure to CO). Exposure to high ambient CO concentrations can result in concentrations of COHb at 2% or higher if the exposure lasts long enough (hours). The exposure time is critical as there is an 8-12 hour period necessary for equilibrium between ambient CO and blood COHb concentration. For people who smoke, cigarettes are usually the most significant source of personal CO exposure. COHb concentrations average about 5% in smokers and are up to 10% or even higher in some very heavy smokers (Beckett 1994).

The CO ambient-air health standards set by EPA are intended to keep COHb concentrations for nonsmokers below 2% and protect the most susceptible members of the population. The goal of both the 1-hour and 8-hour standards is related to maintaining COHb concentrations below this level. Because of the time required for ambient CO to affect COHb levels, it requires exposure to 35 ppm of CO for 1 hour to achieve approximately the same level of COHb as exposure to 9 ppm of CO for 8 hours. EPA (2000a) provides a comprehensive review of the literature pertaining to the health effects of CO for typical environmental exposures that would be associated with COHb levels less than 10%. The major findings presented by EPA (2000a) are summarized below; the committee accepts these

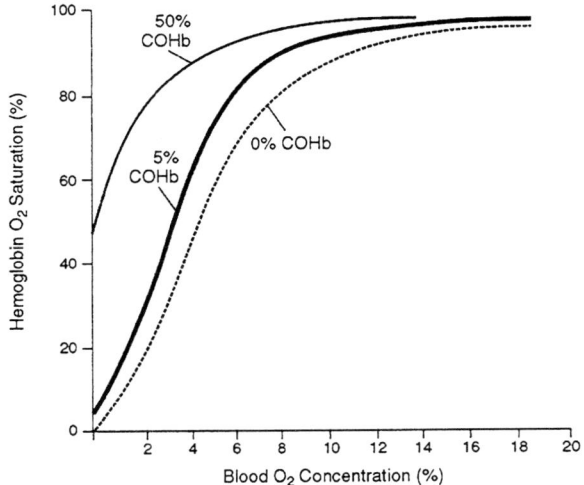

FIGURE 1-6 Diagram of hemoglobin response to the presence of COHb. The concentration of O_2 in the environment surrounding the hemoglobin is shown on the x-axis. The O_2 saturation, or how much of the hemoglobin's capacity for storing O_2 is used, is shown on the y-axis. At higher O_2 concentrations, as are found in the lungs, the hemoglobin can be more O_2 saturated. Likewise, at lower O_2 concentrations, as are found in other parts of the body, O_2 will dissociate from the hemoglobin to achieve O_2 percent saturations as indicated by the curve. The presence of COHb shifts this curve to the left. For a given O_2 concentration, the hemoglobin will require a higher O_2 percent saturation and allow less O_2 to be released to the body's tissues. Source: Adapted from Shephard 1983.

findings as sufficient evidence of the health effects caused by exposure to CO at concentrations of 9 ppm and above for extended periods of time. However, the committee has noted that the public may not be sufficiently aware that CO poses a health threat.

The acute affects of CO poisoning are well understood (Raub et al. 2000). Generally, in otherwise healthy people, headache develops when COHb concentrations reach 10%; tinnitus (ringing in the ear) and lightheadedness at 20%; nausea, vomiting, and weakness at 20-30%; clouding of consciousness and coma at around 35%; and death at around 50% (Coburn 1970). However, the outcomes of long-term, low-concentration CO exposures are not as well understood. Because of the critical nature of blood flow and O_2 delivery to the heart and brain, these organ systems, as well as the lungs (the first organ to come into contact with the pollutant), have received the most attention.

The most well-documented effect in controlled-exposure studies is that of CO on reproducible exercise-induced angina. Angina is a type of chest pain that occurs when there is not enough blood flow to the heart muscle, and it is a symptom of coronary artery disease. In patients with known coronary artery disease, COHb concentrations as low as 3% exacerbate the development of exercise-induced chest pain (Allred et al. 1989a). Concentrations as low as 6% are associated with an increase in the number and frequency of premature ventricular contractions of the heart during exercise in patients with severe heart disease (Allred et al. 1989b; Sheps et al. 1990). These results have provided support for epidemiological studies associating ambient CO with heart-disease exacerbation. Large cohort studies of environmental exposures have confirmed that high daily ambient CO concentrations are associated with statistically significant increases in the numbers of hospital admissions for heart disease (Poloniecki et al. 1997; Schwartz 1999) and congestive heart failure (Morris et al. 1995) and with increases in deaths from cardiopulmonary illnesses (Prescott et al. 1998).

Neuropsychiatric (neurological and psychiatric) disorders and cognitive impairments due to long-term, low-concentration CO exposures have been hypothesized, in part on the basis of extrapolation from the known acute effects of high-dose CO poisoning. In clinical experiments on healthy volunteers, controlled CO exposure was associated with subtle alterations in visual perception when COHb concentrations were above 5% (McFarland 1970; Horvath et al. 1971). However, the significance of this finding remains unknown. Similar studies have shown measurable but small effects on auditory perception, driving performance, and vigilance (Beard and Wertheim 1967; McFarland 1973; Benignus et al. 1977). The neurobehavioral effects described in Beard and Wertheim (1967) served as the scientific basis for the original CO standard promulgated in 1971. However, later studies questioned those results, so the current standard is based on the aggravation of angina pectoris and other cardiovascular diseases.

The role of CO in pulmonary disease is unclear. In the Seattle area, a single-pollutant model showed a 6% increase in the rate of hospital admissions for asthma with each 0.9-ppm increase in ambient CO, but CO increases were concomitant with increases in other air pollutants (Sheppard et al. 1999). In Minneapolis and Toronto, CO concentrations showed weak and inconsistent associations with total admissions for respiratory diseases (Burnett et al. 1997; Moolgavkar et al. 1997). EPA (2000a) cautions that the biological plausibility of CO's association with respiratory illness is tenuous, because the mechanism by which ambient CO exposures could produce or promote harmful respiratory effects has not been demonstrated. As described in subsequent sections, CO is closely associated with copollutants, including hydrocarbons (HCs)[5] and fine particulate matter ($PM_{2.5}$) from motor-vehicle emissions. The respiratory effect attributed to CO might be the result of exposure to HCs (Pappas et al. 2000) or motor-vehicle-related $PM_{2.5}$ (Buckeridge et al. 2002).

A fetus is more susceptible to CO than an adult; the O_2-hemoglobin dissociation curve is to the left of that in the adult and is shifted even further to the left by CO exposure. Also, because the half-life of fetal COHb is longer than that in adults, it may take up to five times longer for its concentrations to return to normal. Studies have shown that exposure to high concentrations of CO during the last trimester of pregnancy may increase the risk of low birth weights and that exposures to CO and airborne particulate matter (PM)[6] during pregnancy may trigger preterm births (Ritz and Yu 1999; Ritz et al. 2000). A recent study correlated CO and O_3 exposure during pregnancy to birth defects such as cleft lip and defective heart valves (Ritz et al. 2002). The correlation of these effects with ambient CO occurred at concentrations below the NAAQS. The study was inconclusive regarding the effects of PM_{10} and nitrogen dioxide; however, the lead author of the study cautioned that the real culprit might be other pollutants, such as PM and some air toxics, that are coemitted with CO in tailpipe emissions (Ritz 2002).

Public-health laws are designed to protect the most susceptible members of the population. People with coronary artery disease or other cardiopulmonary diseases, fetuses, infants, and athletes who exercise heavily in high-CO atmospheres are particularly susceptible to adverse health effects

---

[5]The terms volatile organic compounds (VOCs) and HCs are used to denote organic compounds that are emitted as vapors under typical atmospheric conditions. Unless quoting a source or a regulation that uses another term, the report uses the term HC exclusively. Appendix B describes the differences among the terms used to refer to HCs.

[6]Airborne particulate matter (PM) refers to a broad class of discrete solid particles and liquid droplets of varied chemical composition and size. PM_{10} refers to the subset of PM with an aerodynamic diameter less than or equal to a nominal 10 micrometers. $PM_{2.5}$ refers to the subset of PM with an aerodynamic diameter less than or equal to a nominal 2.5 micrometers.

> **Recommendations: Health Effects**
>
> To reduce the potential adverse health effects of CO, the few remaining areas not in attainment need to continue making progress toward meeting and maintaining the CO standard. Public-health issues associated with ambient CO should be emphasized through enhanced public-awareness campaigns. Further study to reveal the effects of CO on the fetus and to separate the effects of CO from its copollutants is encouraged. Also, there should be more toxicology studies of the automobile exhaust mixture.

from CO. The evidence summarized above, and described more fully in EPA (2000a), indicates that attainment of the ambient-CO standards can decrease morbidity and mortality from atherosclerotic heart disease. Although less conclusive, there is evidence that attainment of the CO standards will also decrease fetal loss and childhood developmental abnormalities. These health benefits translate into economic savings associated with avoided health care and avoided work-time losses as well as intangible savings in quality of life.

Control of CO through new-vehicle emissions standards has also had a significant collateral public-safety benefit through the reduction of accidental CO poisoning (Cobb and Etzel 1991; Shelef 1994; Marr et al. 1998). Mott et al. (2002) recently used computerized death-certificate data maintained by the Centers for Disease Control and Prevention to evaluate the influence of national vehicle emissions controls on unintentional motor-vehicle-related CO deaths between 1968 and 1998. They estimated that over 11,000 deaths were avoided because of these standards, a reduction in unintentional motor-vehicle-related CO mortality from 4.0 to 0.9 deaths per 1 million person-years.

Summary of CO Benefits and Costs from the Clean Air Act

Although there have been no comprehensive assessments of health benefits from controlling CO at individual locations, including meteorological and topographical problem areas, EPA has estimated nationwide benefits attributable to the Clean Air Act in two reports: *Final Report to Congress on the Benefits and Costs of the Clean Air Act, 1970 to 1990* (EPA

1997a) and *Final Report to Congress on the Benefits and Costs of the Clean Air Act, 1990 to 2010* (EPA 1999). However, these documents do not separate the health benefits of CO control from other criteria pollutants. In the 1997 report, the control of PM and CO under the Clean Air Act is estimated to reduce the mean number of hospitalizations for congestive heart failure by 39,000 annually in 1990 compared with a no-control scenario (EPA 1997a). The no-control scenario assumes that no air pollution controls were established beyond those in place prior to the enactment of the 1970 amendments to the Clean Air Act. In the 1999 report, the control of PM, CO, NO_x, sulfur dioxide, and ozone under the CAAA90 is estimated to reduce the mean number of hospitalizations for respiratory aliments and congestive heart failure by 64,000 annually in 2010 compared with a pre-CAAA90 scenario (EPA 1997a). The pre-CAAA90 scenario assumes that no air pollution controls were established beyond those in place prior to the enactment of the CAAA90. It is clear from these reports why the emphasis in air quality management in the United States is on PM and ozone. For example, the 1999 document estimates that controlling PM under the CAAA90 will reduce premature mortality in 2010 by a mean value of 23,000 annually. No reduction in premature mortality is attributed to CO control.

CO Exposure

Exposures in Vehicles

An issue unique to motor vehicles is the proximity of emissions sources to receptors. Automobile air vents can take in exhaust emissions from other vehicles, thereby accumulating CO in the interior compartment. Studies have shown that when CO concentrations near roadways average 3-4 ppm, the average concentration in rider compartments is typically 5 ppm (Akland et al. 1985; Flachsbart et al. 1987). A study released by the California Air Resources Board (Rodes et al. 1998) reported that CO levels were between 2 and 10 times higher inside vehicles than at roadside or fixed monitoring stations due to simple dispersion of the pollutant. Researchers found similarly high concentrations of HCs and toxic compounds such as benzene and 1,3-butadiene. The relationship of these pollutants to CO is discussed further in a later section of this report. Flachsbart (1999) summarizes exposures to mobile-source CO emissions in various microenvironments and shows how congested roadways, street canyons, tunnels, underpasses, drive-up facilities, and parking garages can produce exposures well above ambient conditions.

Relationship of Indoor to Ambient Concentrations

Most people spend a majority of their time indoors; this is particularly true in cold climate areas during winter, when ambient CO concentrations tend to be highest. That leads to the question of the relationship between indoor and outdoor concentrations. Air pollution in buildings can come from indoor sources and from air exchange with outdoor ambient pollution. Air exchange may be active, as in the case of a mechanical ventilation system, or passive, as in the case of infiltration associated with temperature or pressure differences between the outside and the inside of a building. Though homes in northern climates may be tighter, air exchange through leaks is controlled by the temperature difference, with a large temperature gradient producing a greater infiltration rate. Thus, CO penetrates freely with infiltration air from the outside, even in winter in Fairbanks, and is not removed by building materials or ventilation systems. Furthermore, there are no effective indoor chemical or physical processes for lowering CO on the time scales of interest for exposure and toxic effects. Hence, being indoors offers little protection from outside CO levels.

The relationship between indoor and outdoor CO concentrations can be evaluated with a simple differential mass-balance model (Shair and Heitner 1974) that has the following steady-state solution when we combine active ventilation and passive infiltration into a single air-exchange term:

$$C_i = \frac{paC_o}{a+k} + \frac{S}{(a+k)V}$$

where

C_i = indoor concentration, mg/m³;
C_o = outdoor concentration, mg/m³;
p = penetration coefficient, 0-1;
a = air exchange rate, hour⁻¹;
k = decay rate, hour⁻¹;
S = mass flux of the indoor source, mg/h; and
V = building volume, m³.

For CO, the relationship is simpler because the penetration coefficient (p) is unity and the decay rate (k) is effectively zero. Therefore, the solution is

$$C_i = C_o + \frac{S}{aV}.$$

In the absence of indoor sources (S), the steady-state indoor concentration of CO will equal the average outdoor concentration. When a source of CO is present indoors (e.g., from a faulty furnace, an underground parking garage, a kerosene heater, or a tobacco smoker), the indoor source *adds* to the background concentration from the outdoor air (EPA 2000a). Therefore, buildings do not provide protection from high outdoor concentrations of CO. The idea that buildings provide protection from high outdoor CO concentrations is a common misconception.[7]

Other Exposures

The contribution to personal exposures from certain sources, such as gasoline-powered lawnmowers, snowmobiles, recreation boats, generators, and garden equipment, can be substantially greater than fixed site monitoring data suggest. These sources contribute considerably less to the regional inventory than do mobile sources, and the exposed operators of lawn and garden equipment are working in close proximity to the CO emissions source. In combination with urban background concentrations, localized sources may subject some individuals to very high CO concentrations.

Spatial Distribution of CO

Studies that demonstrate the spatial and temporal distribution of CO are beneficial in assessing the potential human exposure to CO and other pollutants from vehicle emissions. Saturation studies are one method. They typically rely on portable monitors that "saturate" a geographical area with samplers to assess the air quality in places where high concentrations of

---

[7]In this regard, CO is different from O_3, which is highly reactive and is destroyed when infiltrating inside from outdoors.

pollutants are possible. Monitors can be deployed at temporary fixed-site locations or in mobile sampling vehicles. These studies are helpful to air pollution control agencies for evaluating their ambient air monitoring networks, characterizing pollutant concentrations over the entire saturation study area, and locating hot spots or high pollutant impact points. Personal and indoor monitoring could be incorporated into such studies to relate ambient concentrations to personal exposure.

Saturation studies typically involve deploying temporary mobile or stationary monitors throughout a wide study area to characterize the spatial extent of CO. The committee considered several of these studies (Morris 2001; Guay 2001; Lawson 2002; Ransel 2002) during its deliberations. On particular example is the study that was carried out in the Las Vegas Valley during the winter of 2001-2002 (Ransel 2002). The objectives of the study were: (1) to evaluate the adequacy of the monitoring network to measure the spatial distribution of CO and (2) to ensure that no areas of higher concentrations were missed with the existing network. The study collected data from 64 temporary fixed monitoring sites operating continuously for six weeks and from 11 episodes using a mobile sampling van that collected 1-minute average CO concentrations. These measurements were in addition to those made at 14 already existing permanent monitoring stations. The mobile sampling van was equipped with two samplers capable of measuring average CO concentrations and a global positioning system (GPS) to determine location and provide real-time mapping and display. Figure 1-7 shows the locations of CO measurements made with the van. The study concluded that current permanent monitoring sites are suitably located to identify peak CO concentrations and to map areas where relatively higher CO levels may occur.

Another saturation study reviewed by the committee was performed during the winter of 1997-1998 in Anchorage, Alaska (Morris and Taylor 1998; Morris 2001). The objective of the study was to determine whether the permanent monitoring network adequately characterized CO exposures in neighborhoods, near major roadways, and in parking lots. The study used 16 temporary fixed monitors to supplement the 4 permanent monitors. The study concluded that the current permanent network adequately characterizes CO concentrations at roadway sites but might not characterize the upper range of CO concentrations in neighborhoods. The high CO concentrations observed at most residential monitoring sites, often in the morning, indicated that cold-start and/or warm-up idling of vehicles by commuters is a significant source of CO in those locations.

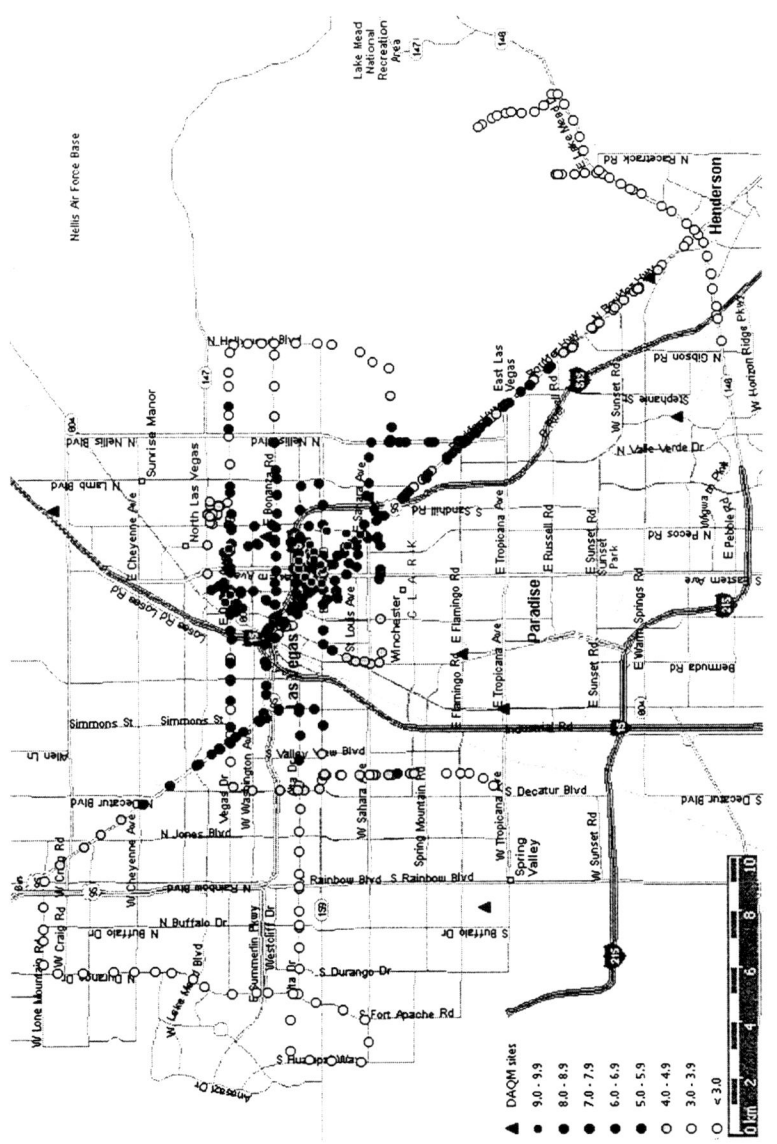

FIGURE 1-7 Locations of 1-min average CO concentration measurements made using a mobile sampling van in the Las Vegas Valley between November 20, 2001, and January 6, 2002. Source: Ransel 2002.

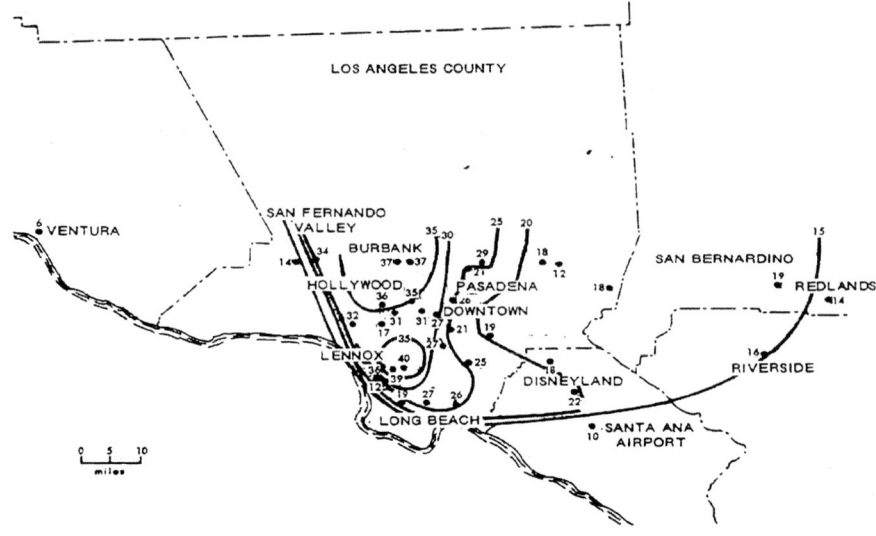

FIGURE 1-8 Maximum 8-hour average CO concentrations in the South Coast Air Basin for 1956-1967. Source: DHEW 1970.

Finally, a comparison of the change in the spatial distribution of the maximum 8-hour CO concentration over time in the Los Angeles area provides a qualitative description of the reduction in exposure to high CO. Figure 1-8 shows that the 8-hour CO concentration was exceeded 0.1% of the time averaged over the period from 1956 through 1967. Since there are 1,095 discrete 8-hour time periods in a year, the 0.1 percentile value approximates the 8-hour concentration likely to be exceeded an average of once per year (DHEW 1970). Figure 1-9 shows the maximum 8-hour CO concentration for 2000 (SCAQMD 2000a). Although 2000 may have been a favorable year in terms of meteorology, it is clear that the 8-hour peak concentrations and the spatial extent of CO pollution have decreased greatly since the 1956-1967 time period.

Roadway Health Effects

The correlation between CO and other motor-vehicle-related emissions is important because of studies linking health impacts and proximity to

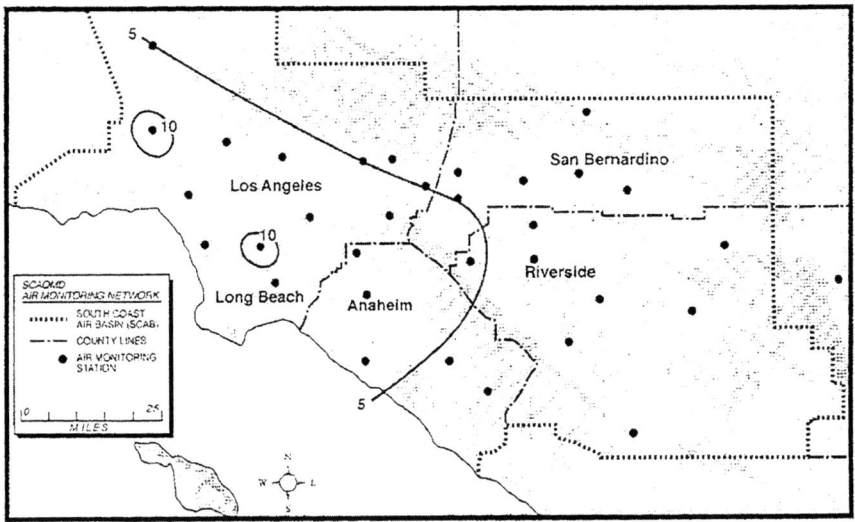

FIGURE 1-9 Maximum 8-hour average CO concentrations in the South Coast Air Basin in 2000. Source: SCAQMD 2000a.

major roadways. CO is a relatively easy pollutant to measure and thus can be an indicator of roadway emissions. As shown in Figure 1-10, CO is highly correlated with black carbon and ultrafine particles in close proximity to highways (Zhu et al. 2002). Brunekreef (1997) found a reduction in lung function in children living near a major highway; Hoek et al. (2002) found an association between cardiopulmonary mortality and proximity to a major roadway; and Buckeridge et al. (2002) reported the effects of motor-vehicle emissions on respiratory health. These studies attribute most health impacts to PM and HCs. Ritz et al. (2002) showed a correlation between traffic-related CO emissions and birth defects in southern California. Recently, Wilhelm and Ritz (2003) reported an association between residential proximity to traffic and adverse birth outcomes such as low birth weight and preterm birth.

RELATIONSHIP OF CO TO OTHER AIR POLLUTANTS

As discussed earlier in this chapter, mobile sources, both on-road and off-road contribute 75-95% of CO emissions in selected urban areas. There-

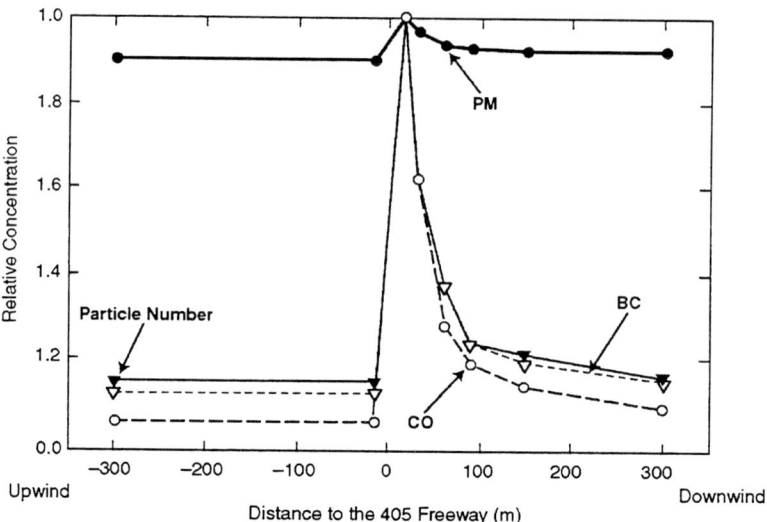

FIGURE 1-10 Relative mass, total particle number, black carbon, and CO concentrations versus downwind distance from a freeway. Source: Zhu et al. 2002. Reprinted with permission; copyright 2002, Air & Waste Management Association.

fore, CO may be an indicator of other, less well-characterized pollutants emitted from vehicles, such as fine particulate matter ($PM_{2.5}$) and air toxics associated with HCs. However, CO has some substantial shortcomings as an indicator of other mobile-source emissions. The correlation between CO and other pollutants, and CO's role in tropospheric ozone, are discussed below. The correlation between CO and other motor-vehicle-related emissions is important because of the studies linking health impacts of air pollution to proximity to major roadways. Those studies are also discussed in this section.

Association of CO Emissions to Other Emissions

Automobile exhaust is a complex mixture of compounds, some of which are classified as criteria air pollutants and others as hazardous air pollutants (HAPs) or "air toxics." The correlation of CO with $PM_{2.5}$ and some air toxics is especially strong for gasoline-powered light-duty vehicles (LDVs) operating under fuel-rich conditions. As discussed more exten-

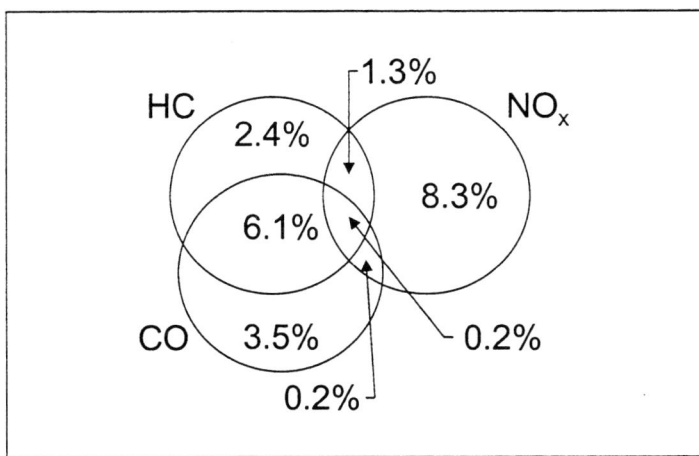

FIGURE 1-11 Degree of overlap among the highest 10% of emitters of CO, HC, and NO_x in the light-duty vehicle fleet, based on the results of emissions tests on 12,977 vehicles administered during random roadside inspections in California, from June 9, 1998, until October 29, 1999. Note that the sizes of the overlapping areas are not drawn to scale. Of the vehicles tested, 78% did not fall in the top 10% for CO, HC, or NO_x. Source: Diagram prepared by Gregory S. Noblet, University of California, Berkeley. Reprinted with permission.

sively in Chapter 3, fuel-rich conditions exist when excess fuel is introduced into the engine combustion process, greatly increasing the production of CO, unburned HCs, and $PM_{2.5}$. Virtually all gasoline-powered vehicles are designed to operate under fuel-rich conditions during cold-start operation, leading to a significant proportion of total emissions. In Fairbanks, Alaska, winter cold-start and initial-idle emissions contributed an estimated 45% of overall on-road emissions (NRC 2002). Fuel-rich conditions also occur during hard accelerations and climbing up grade, when the fuel-metering system injects extra fuel to improve vehicle performance, or because of malfunctions in fuel-metering and other system components.

Substantial evidence demonstrates that a large fraction of emissions is from a relatively small percentage of LDVs[8] that have a disproportionate impact on total air pollution from mobile sources. Figure 1-11 (above) shows the overlap among the highest-emitting 10% of vehicles randomly

---

[8]Typical numbers reported in the literature (usually obtained from measurements of in-use vehicles) show that 50-60% of on-road LDV exhaust emissions are produced by about 10% of LDVs (NRC 2001).

pulled over and tested for CO, HCs, and nitrogen oxides (NO_x) in California. The figure indicates that there are significant similarities in the high-emitting subset of vehicles, especially regarding CO and HCs.

EPA lists 21 mobile-source (on-road and nonroad) air toxics, shown in Table 1-4. For some of these (e.g., arsenic and dioxin), emissions inventories show that mobile sources contribute only a small fraction to their overall emissions, but for most, mobile sources are significant if not dominant among contributors. However, it should be noted that the uncertainties associated with emissions inventories for air toxics and PM are likely greater than for CO and HCs. Inventories for these pollutants piggyback on estimates for CO and HC, introducing another level of uncertainty. In addition, less ambient monitoring and emissions data are available to develop and evaluate these emissions. Table 1-5 lists direct emissions by source category for five important mobile-source air toxics. All of these air toxics are either known or probable human carcinogens, and some have additional noncancer health effects. The top four show a sizable fraction of emissions associated with LDVs, which are the largest source of CO emissions. However, both formaldehyde and acetaldehyde have significant secondary sources (from atmospheric chemical reactions of VOCs, including VOCs emitted from LDVs) as well as direct emissions that contribute to their ambient concentrations.

The relationship between CO and $PM_{2.5}$ emissions is highly uncertain. Ambient observations in urban areas tend to show that a significant fraction of $PM_{2.5}$ emissions come from mobile sources (NARSTO 2003). PM diesel emissions have garnered particular attention and have been classified as an air toxic (EPA 2002c). The Multiple Air Toxics Exposure (MATES-II) study in the South Coast Air Basin of California (SCAQMD 2000b) estimated that PM diesel emissions have a much higher cancer risk compared with all other air toxics combined. PM diesel emissions are primarily from heavy-duty diesel vehicles (HDDVs) and off-road diesel engines, which tend to have very low CO emissions compared with LDVs. The emissions from HDDVs and LDVs are correlated because both sources travel the same roadways, but the spatial and temporal patterns of the emissions from these two vehicle classes may differ greatly. For example, Lena et al. (2002) found that site-to-site variability in the number of large trucks[9] was much greater than that for light-duty vehicles, and that the ratio of passen-

---

[9]Lena et al. (2002) defined large trucks as those with two axles that had four tires on the rear axle, or trucks with more than two axles.

TABLE 1-4 Mobile-Source (On-Road and Nonroad) Air Toxics Identified by EPA and the Percent of National Emissions from Mobile Sources

Air Toxic	Mobile-Source Emissions (%)	Air Toxic	Mobile-Source Emissions (%)
Acetaldehyde	70	Lead compounds	23
Acrolein	39	Manganese compounds	1.5
Arsenic compounds	0.6	Mercury compounds	4
Benzene	76	Methyl *tertiary*-butyl ether (MTBE)	86
1,3-Butadiene	60		
Chromium compounds	4.2	Naphthalene	unknown
Dioxin/furans	0.2	Nickel compounds	8.5
HDDV diesel particulate matter and diesel exhaust organic gases	100	Polycyclic organic matter (POM)	6
		Styrene	40
Ethylbenzene	84	Toluene	74
Formaldehyde	49	Xylene	79
n–Hexane	44		

Source: EPA 2000c.

ger cars to large trucks varied greatly by site. Further, CO-related regulations will have little impact on diesel PM concentrations.

The contribution of LDV emissions to $PM_{2.5}$ concentrations is an area of active research. A study of $PM_{2.5}$ in Denver, Colorado, found that LDVs contributed a much larger fraction of $PM_{2.5}$ emissions than did diesel vehicles, although it is not clear whether that result is unique to the location (Fujita et al. 1998; Norton et al. 1998). This is in contrast to a study from southern California that found diesels to be the dominant contributor of mobile-source-emitted $PM_{2.5}$ (Schauer et al. 1996). ARCADIS G&M (2003) recently studied seven cities and found that LDVs were the dominate contributor to mobile-source-emitted $PM_{2.5}$ in Birmingham, Alabama, and Westbury, Connecticut, and they contributed approximately the same amount as diesels in Las Vegas, Nevada. This study found diesels to be the dominant source in Albany, New York; Houston, Texas; Long Beach, California; and El Paso, Texas. In terms of toxicity, a recent study found no difference in the toxicity of particles emitted from diesel and LDVs, but particles from diesel and LDV high-emitters were much more potent on an equivalent mass than those from normal-emitters (Seagrave et al. 2002).

TABLE 1-5 Direct Emissions of Selected Mobile-Source Air Toxics

	Total 1996 Emissions (short tons)	Mobile-Source Contribution			Mobile-Source Contribution
Air Toxic		LDV (short tons)	HDV (short tons)	Nonroad Sources	
Benzene	351,000	155,000 (44%)	13,000 (4%)	99,000 (28%)	267,000 (76%)
Formaldehyde	346,000	53,000 (15%)	30,000 (9%)	86,000 (25%)	169,000 (49%)
Acetaldehyde	99,000	19,000 (19%)	9,000 (9%)	41,000 (41%)	70,000 (70%)
1,3-Butadiene	56,000	20,000 (36%)	4,000 (7%)	10,000 (18%)	33,000 (60%)
Diesel PM	524,000			341,000 (65%)	524,000 (100%)
$PM_{2.5}$	8,194,000	65,000	155,000	410,000	631,000

Note: Secondary sources for some of these pollutants (formaldehyde, acetaldehyde, $PM_{2.5}$) are significant.
Sources: EPA 2000b,c; EPA 2001c; EPA 2003a.

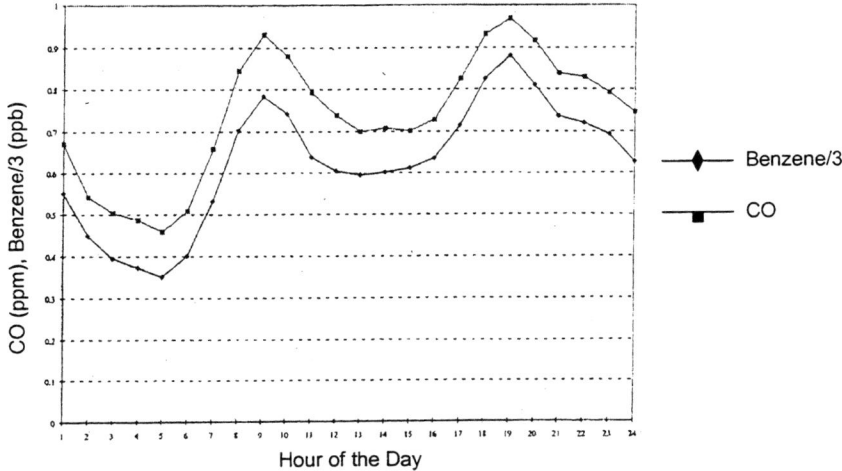

FIGURE 1-12 Diurnal average CO and benzene in London Bloomsbury in 1996. Source: Williams 2000. Reprinted with permission; copyright 2000, Elsevier SAS.

Ambient CO Concentrations and Other Pollutants

The relationships among ambient concentrations of CO, air toxics, and PM are complex and are affected by differences in direct pollutant sources and by atmospheric processes that create chemical sinks and secondary products. Figure 1-12 (above) indicates that benzene and CO concentrations have similar diurnal patterns. Benzene concentrations also show a seasonal pattern similar to that of CO, with maximum concentrations occurring during winter. The correlation between ambient CO and benzene concentrations stems from a similarity in emissions sources and benzene's fairly long atmospheric lifetime, which allows it to be dispersed with CO.[10] Ambient measurements in the Los Angeles area have shown strong correlations between ambient levels of benzene and CO ($r^2 = 0.76$) (Figure 1-13), and an even stronger correlation was observed between CO and ambient levels of the relatively short-lived species 1,3-butadiene ($r^2 = 0.84$) (Figure

---

[10]EPA (2002c) estimates that the atmospheric lifetime of benzene is 11 days. That means that benzene is removed from the urban environment by meteorological processes (as opposed to a chemical sink), which is how most CO is removed.

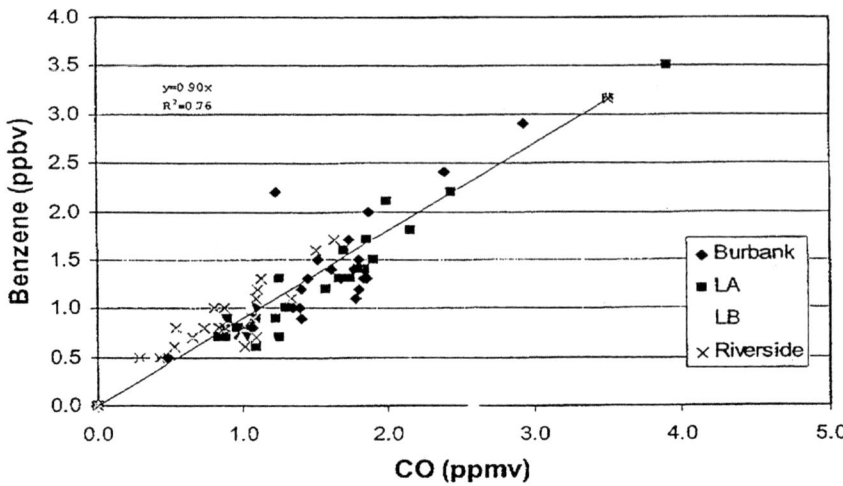

FIGURE 1-13 Benzene versus CO for four sites (Burbank, Los Angeles, Long Beach, and Riverside) in California's South Coast Air Basin in 1996. Source: CARB 1999.

1-14) (CARB 1999). MATES-II found that benzene, 1,3-butadiene, and other air toxics (methylene chloride, perchloroethylene, lead, and elemental carbon) have seasonal concentrations that peak during late fall and winter in the South Coast Air Basin (SCAQMD 2000b). This was ascribed to local seasonal meteorological conditions—light winds and surface inversions inhibiting vertical dispersion of pollutants. Table 1-6 shows atmospheric lifetimes for selected air toxics.

Formaldehyde and acetaldehyde are reactive in the atmosphere. They have lifetimes of a few hours during daylight (Atkinson 2000). Secondary emissions sources greatly influence the concentrations of these compounds. The chemical reactions that lead to the formation of additional formaldehyde and acetaldehyde from other HCs depend on solar radiation; therefore, higher concentrations occur during months with greater solar radiation. The MATES-II study (SCAQMD 2000b) found that these concentrations peak in the summer and fall in the South Coast Air Basin. The peak is delayed because increased vertical mixing and dispersion occurs during the summer, which reduces the concentrations of these pollutants.

Ambient concentrations of $PM_{2.5}$ are influenced by widely varying emissions and atmospheric processes. A recent assessment found that Mexico City and many areas in the western United States have their highest

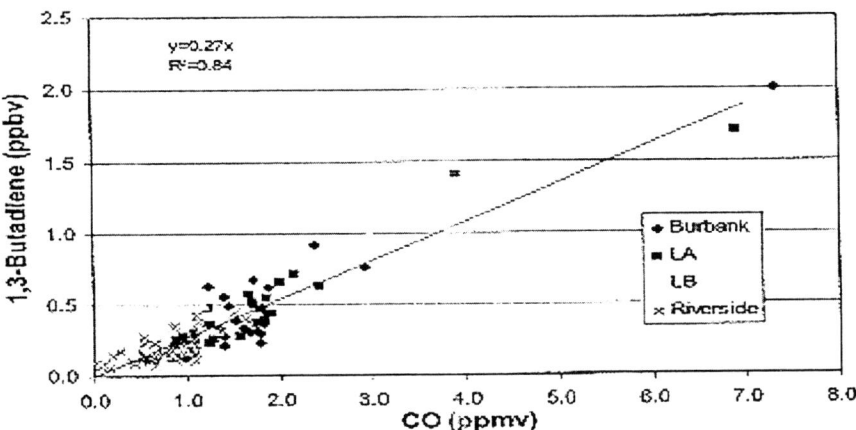

FIGURE 1-14 1,3-Butadiene versus CO for four sites (Burbank, Los Angeles, Long Beach, and Riverside) in California's South Coast Air Basin in 1996. Source: CARB 1999.

concentrations of PM during winter because of limited dispersion during winter months (NARSTO 2003).[11] Figure 1-15 shows the correlation between daily average CO and $PM_{2.5}$ concentrations for the winter of 2000-2001 in Fairbanks, Alaska. This limited data set shows a correlation coefficient (R-squared) of 0.70. The meteorological conditions that lead to CO buildup may also play a role in episodes of high PM. Thus, cities that have had problems coming into compliance with the 8-hour NAAQS for CO because of their meteorology and topography may also be susceptible to violations of the 24-hour NAAQS for $PM_{2.5}$. However, Figure 1-15 makes clear that high CO levels do not necessarily coincide with high levels of $PM_{2.5}$. Changes in emissions-producing activities, for example, a weekend when more people are home using a fireplace instead of out driving, might shift the result of an inversion episode from high CO to relatively higher concentrations of $PM_{2.5}$. In addition, the meteorological conditions that produce CO exceedances may be slightly different from those that produce high $PM_{2.5}$ concentrations. Also, secondary formation of $PM_{2.5}$ (from gas-to-particle conversion of nitrates, sulfates, and organics)

---

[11]This is in contrast to most areas in the eastern U.S. that have peak PM concentrations in the summer.

TABLE 1-6 Calculated Atmospheric Lifetimes for Selected Compounds

Compound	Atmospheric Lifetime[a] (daylight hours)
Acetaldehyde	5.9
Benzene	75.3
1,3-Butadiene	1.4
Carbon monoxide	440.9
Formaldehyde	9.9

[a]The atmospheric lifetime for each compound was calculated based on the recommended OH rate constants and a 12-hour average OH radical concentration of 3.0×10^6 molecule/cm^3.
Source: Atkinson 1994.

can occur, especially during summer when photochemistry is most prevalent, and further obscure the relationship between CO and $PM_{2.5}$.

For example, Salt Lake City, Utah, was originally in nonattainment for the CO NAAQS. In addition, the area had exceedances[12] of the 24-hour $PM_{2.5}$ standard of 65 micrograms/cubic meter four times: January 1, 2000, at the Cottonwood site (24-hour value = 71.3 µg/m^3); December 30 and 31, 2000, at the Hawthorne site (24-hour values = 72.4 and 66.3 µg/m^3, respectively), and December 30, 2000 at the North Salt Lake site (24-hour value = 68.7 µg/m^3). CO measurements were not particularly high on those days. The highest 8-hour average CO concentration at the Cottonwood site on January 1, 2000, was 2.7 ppm. The highest 8-hour average CO concentration at Hawthorne on December 30, 2000, was 3.4 ppm, and on January 31 it was 2.0 ppm. (There is no CO analyzer located at the North Salt Lake site.) According to Robert Dalley of the Utah Department of Environmental Quality (personal communication, September 20, 2002), high $PM_{2.5}$ and high CO concentrations occur in response to prolonged winter temperature inversions. The inversions can last 2 to 3 weeks without a break, as was the case during these $PM_{2.5}$ exceedances. When inversions have some fog associated with them, $PM_{2.5}$ values are high, but CO values remain relatively low. When clear skies accompany inversions, $PM_{2.5}$ concentrations

---

[12]The $PM_{2.5}$ standard is written as an 95th-percentile exceedance; thus, a single exceedance of this standard does not mean the area is out of attainment.

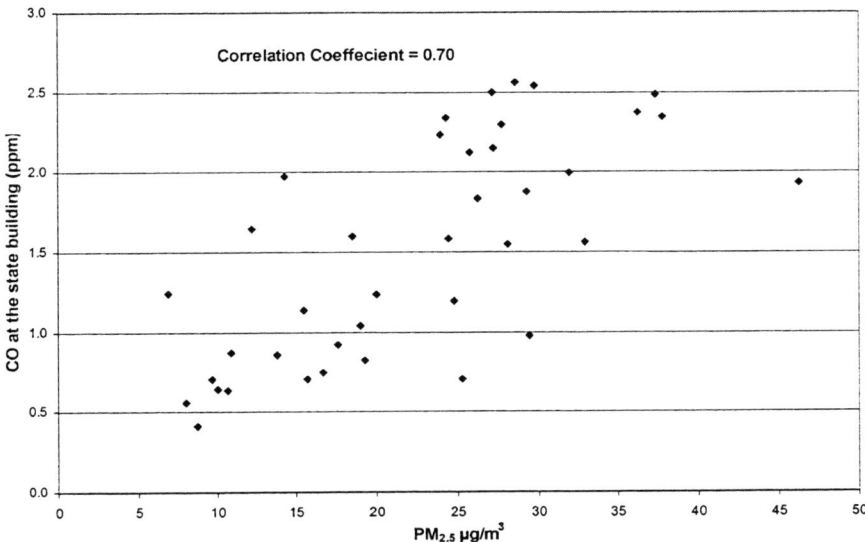

FIGURE 1-15 Correlation of daily average CO and $PM_{2.5}$ concentrations at the state building, Fairbanks, Alaska, November 2000 to February 2001.

are lower, and CO concentrations are high. A hypothesis explaining these observations might be that during inversions with fog, aqueous reactions in the fog form secondary $PM_{2.5}$ more quickly, and during clear inversions stratification in the inversion traps CO closer to the ground. However, this conclusion has not been confirmed.

Roles of CO in Tropospheric Ozone and Climate Change

Tropospheric Ozone

In the atmosphere, the only chemical loss process for CO is by reaction with the hydroxyl (OH) radical. The overall reaction is

$$OH + CO\ (+ O_2) = HO_2 + CO_2.$$

Using a global average tropospheric OH radical concentration of 9.4×10^5 molecule cm^{-3} (Prinn et al. 2001), the average CO lifetime is calculated to be 2 months, which is sufficiently long for CO emitted in the United States to be mixed throughout the northern hemisphere.

When enough NO is present that HO_2 radicals react only with it,

$$HO_2 + NO = OH + NO_2,$$

the photolysis of NO_2 and the rapid reaction of the oxygen, $O(^3P)$, atom with O_2,

$$NO_2 + \text{sunlight} = NO + O(^3P)$$
$$O(^3P) + O_2 + M = O_3 + M \;(M = \text{air}),$$

leads to net formation of ozone from the reaction of OH radicals with CO (in the presence of NO such that HO_2 radicals react dominantly with NO),

$$CO + 2O_2 = CO_2 + O_3.$$

Therefore, CO can be viewed as the simplest ozone-forming "hydrocarbon."

Because CO reacts rather slowly in the atmosphere, and its photo-oxidation results in the conversion of only one molecule of NO to NO_2 per molecule of CO oxidized (see above), CO has a significantly lower ozone-forming potential (grams of O_3 formed per grams of reactant emitted) than the HC mix in vehicle exhaust (Carter 1998). The 1997 maximum incremental reactivity (MIR) scale (Carter 1998) of CO is 0.065 g of O_3 per gram of CO emitted, which can be compared to the MIR of exhaust emissions from vehicles fueled with two gasolines representative of California reformulated gasoline (testing conducted during the Auto/Oil Air Quality Improvement Research Program) of approximately 3.5 g of O_3 per gram of HC emitted (NRC 1999). However, because of the amount of CO and HCs emitted in vehicle exhaust, NRC (1999) concluded that CO from LDVs contributes 15-25% of the total ozone-forming potential of exhaust emissions. Therefore, despite its low ozone-forming potential, CO contributes to ozone formation in polluted atmospheres. Keep in mind however that ozone formation requires sunlight and is strongly temperature dependent, so it tends to be more of a problem during the summer months and not during the winter months when CO exposure tends to be a problem.

Climate Change

CO contributes to climate change in four ways: (1) it is itself a greenhouse gas (GHG), though its warming potential is much less than that of CO_2; (2) CO is oxidized to CO_2, as shown above, noting that direct emissions of CO from vehicles are an order of magnitude or more lower than those of CO_2; (3) at low NO_x concentrations, reaction of CO with OH radicals results in loss of OH radicals, and by removing OH from the atmosphere CO tends to increase the lifetime of methane (CH_4), a powerful GHG; and (4) CO contributes to the formation of ozone (O_3), another GHG.

Using CO As an Indicator of Other Pollutants

In urban environments, CO can serve as an indicator of motor vehicle emissions from gasoline-fueled vehicles. The observed spatial and temporal variability of CO shows that the effects of motor-vehicle pollution are heterogeneously distributed in urban areas and that CO can be a useful gauge of long-term human exposure to other pollutants of concern, including certain mobile-source air toxics. CO levels also demonstrate the existence of "hot spots" in urban environments where high concentrations of CO and other products of automobile exhaust occur.

However, CO is not a perfect indicator. CO does not react on the time scales of concern for urban pollution, and it is not representative of the chemical reactivity of other pollutants. The weather conditions that produce high CO concentrations are generally unrelated to those that produce ozone pollution, which is most severe during the summer months. Because CO pollution is primarily due to exhaust emissions from LDVs in the urban environment, it is not strongly correlated with evaporative toxic emissions, diesel PM emissions, or stationary- and area-source emissions.

Because CO is formed with other products of incomplete combustion (including unburnt fuel), CO emissions from a specific vehicle often correlate with emissions of HCs and organic compounds (including the air toxics formaldehyde, acetaldehyde, 1,3-butadiene, and benzene). Emissions of formaldehyde and acetaldehyde depend on the fuel used and, more specifically, on the presence of MTBE or ethanol in the fuel. MTBE leads to higher emissions of formaldehyde, and ethanol leads to higher emissions of acetaldehyde (NRC 1999). The exhaust emissions of benzene and other

> **Recommendations: CO As an Indicator of Motor-Vehicle Pollutants**
>
> The committee has several recommendations with regards to the use of CO to represent the distribution of other pollutants. CO can be used to demonstrate the spatial distribution of some mobile-source pollutants, to identify hot spots, and to improve model representation of relationships between transportation activity and emissions. CO can also be used to approximate the concentrations of some air toxics arising from motor-vehicle exhaust emissions, such as benzene and 1,3-butadiene, and perhaps directly-emitted $PM_{2.5}$. CO is most useful as an indicator in the microscale setting where concentrations of pollutants vary dramatically over short distances (e.g., with distance from a roadway). It is less reliable in representing regional distributions of these pollutants and is probably a poor indicator of motor-vehicle air toxics, such as formaldehyde and acetaldehyde, that react rapidly and have substantial sources in the atmosphere.

higher emissions of formaldehyde, and ethanol leads to higher emissions of acetaldehyde (NRC 1999). The exhaust emissions of benzene and other aromatic HCs depend on the fuel aromatic content, because benzene and other aromatic HCs are emitted as unburnt fuel and are formed in the combustion process.

The precise correspondence of CO and other organic vehicle emissions depends on the fuel used (including the presence or absence of a fuel oxygenate, the actual oxygenate used, and the aromatic content of the fuel) and onboard emissions control technologies and engine conditions (i.e., cold start, hard acceleration, etc.). Furthermore, CO's relationship with other organic emissions varies as a function of time after emission. While CO is essentially nonreactive on day-long time scales, most other organic compounds in vehicle exhaust are significantly more reactive than CO, and certain organic compounds (e.g., carbonyl compounds, alkyl nitrates, and peroxyacyl nitrates) are later formed in the atmosphere from atmospheric reactions of other HCs (Atkinson 2000). For example, formaldehyde is removed rapidly by photolysis (and less rapidly by reaction with OH radicals) and has a lifetime of about 4 hours in overhead sun. It is also formed in the atmosphere from the photooxidation of almost all other HCs (Atkinson 2000).

Therefore, although CO can serve as a general indicator of motor-vehicle exhaust (and hence of exposure to vehicle exhaust and/or to photochemically processed vehicle exhaust), CO concentrations alone are uncertain estimators for concentrations of other organic compounds in the same air mass, except for other long-lived vehicle exhaust components such as benzene (and even then, only in the absence of other sources). However, reasonably strong correlations between CO and the shorter-lived volatile organic compounds (VOCs) emitted from LDVs will still be observed over distance scales corresponding to travel times of the pollutants of approximately a half-life or less (see Figure 1-14).

EQUITY CONSIDERATIONS IN THE SPATIAL DISTRIBUTION OF AMBIENT CO

Because CO levels are not evenly distributed, exposure to CO within the population will vary. Individuals living in or near areas of high CO ("hot spots") are exposed to higher concentrations of CO and other mobile-source-related pollutants. Although the network of CO monitors is too sparse to identify all hot spots, the characteristics of the residents living near known hot spots can be examined. An analysis of data from the 2000 U.S. Census shows that the individual CO monitors that registered exceedances of the 9-ppm 8-hour average CO standard during the period 1995-2001 are often found in areas that have greater percentages of low-income and minority residents than their surrounding regions (Table 1-7). All but six of the monitor areas, as defined by the census tract or tracts immediately surrounding each CO monitor, had higher percentages of nonwhite residents in 2000 than the region as a whole. In the area around the Sunrise Avenue monitor in Las Vegas, for example, 49.6% of residents are nonwhite, compared with 26.2% for the Las Vegas Metropolitan Statistical Area (MSA). Three monitor sites in Los Angeles have a lower percentage of nonwhite residents than the region as a whole, but even in those areas, over one-third of residents are nonwhite. The percentage of residents who are of Hispanic origin is higher than the regional share for all but three monitors. The differences are often dramatic. The population in the Las Vegas-Sunrise Avenue area is 68.7% Hispanic compared with 5.3% for the region, and percentages of Hispanic residents for the Phoenix monitor areas are over twice the percentage for the region. Per capita incomes were lower for residents in the monitor area than on average for the region (for all but four of the monitor areas) and were less than half of the regional average for the Las Vegas-Sunrise Avenue, Phoenix-Indian School Road, and Lynwood-Long Beach Avenue monitor areas. The monitor area in Denver at Speer and Auraria Parkway, where the per capita income is over twice

TABLE 1-7 Population Characteristics in Monitor Area[a] Versus Region,[b] 2000

Monitor (ID)	Exceedance Days 1995-2001	% Non-White		% Hispanic		Per Capita Income		% Non-Driving[c]		Population Per Square Mile	
		Monitor Area	Region	Monitor Area	Region	Monitor Area	Region	Monitor Area	Region	Monitor Area	City[d]
Birmingham - Shuttlesworth and 41st[e] (01-073-6004)	69	96.9	32.7	1.2	1.8	8,085	21,142	28.7	4.8	511	595
Calexico - 129 Ethel St. (06-025-0005)	59	51.0	50.6	93.9	72.2	10,193	13,239	16.0	10.3	5,441	4,353
Lynwood - Long Beach (06-037-1301)	58	65.3	51.3	87.0	44.6	7,739	20,683	12.7	14.6	17,827	14,389
Fairbanks - Cushman (02-090-0002)	16	31.2	22.2	4.8	4.2	20,921	21,553	32.6	10.4	3,042	949
Fairbanks - Gilliam Way (02-090-0020)	13	48.6	22.2	5.6	4.2	15,886	21,553	13.4	10.4	3,547	949
Fairbanks - 7th Ave. (02-090-0013)	7	31.2	22.2	4.8	4.2	20,921	21,553	32.6	10.4	3,042	949
Phoenix - Grand Ave. and Thomas Rd. (04-013-0022)	7	38.5	23.0	62.5	25.1	13,109	21,907	18.6	10.0	6,117	2,782

Hawthorne - 120th St. (06-037-5001)	7	46.4	51.3	47.6	44.6	21,148	20,683	8.2	14.6	7,679	13,879
Spokane - Third Ave. (53-063-0044)	5	12.7	8.6	3.2	2.8	19,016	19,233	43.5	11.0	4,798	3,387
Burbank - W. Palm Ave. (06-037-1002)	4	34.6	51.3	33.2	44.6	20,275	20,683	10.3	14.6	11,966	5,782
Las Vegas - East Charleston (32-003-0557)	4	34.2	26.2	32.6	5.3	15,935	21,697	13.1	9.8	8,609	4,223
Las Vegas - Sunrise Ave. (32-003-0561)	3	49.6	26.2	68.7	5.3	10,413	21,697	22.9	9.8	11,878	4,223
Reseda - Gault St. (06-037-1201)	3	41.2	51.3	45.4	44.6	15,069	20,683	16.6	14.6	11,444	2,344
Anchorage - 3201 New Seward Hwy (02-020-0037)	3	33.6	27.8	7.1	5.7	26,260	25,287	17.4	11.0	4,297	153
El Paso - North Campbell (48-141-0027)	3	17.4	26.1	93.6	78.2	3,907	13,139	19.9	7.9	4,519	2,263
Denver - Broadway - Camp (08-031-0002)	2	48.6	20.6	40.1	18.8	20,300	26,206	47.2	10.1	6,041	3,617
Denver - Speer and Auraria Pkwy (08-031-0019)	2	19.5	20.6	9.2	18.8	68,944	26,206	55.9	10.1	5,139	3,617

(Cont.)

TABLE 1-7 Continued

Monitor (ID)	Exceedance Days 1995-2001	% Non-White Monitor Area	% Non-White Region	% Hispanic Monitor Area	% Hispanic Region	Per Capita Income Monitor Area	Per Capita Income Region	% Non-Driving[c] Monitor Area	% Non-Driving[c] Region	Population Per Square Mile Monitor Area	Population Per Square Mile City[d]
Kalispell - Idaho and Main (30-029-0045)	2	3.5	3.7	1.3	1.4	19,085	17,915	13.0	10.8	2,047	2,606
Anchorage - 3201 Turnagain (02-020-0048)	2	32.6	27.8	6.5	5.7	23,388	25,287	31.0	11.0	7,192	153
Spokane - Hamilton St. (53-063-0040)	1	11.7	8.6	4.6	2.8	10,838	19,233	26.3	11.0	5,913	3,387
Phoenix - Indian School Rd. (04-013-0016)	1	48.8	23.0	60.5	25.1	9,986	21,907	11.0	10.0	8,776	2,782
Provo - 242 N. University Ave. (49-049-0004)	1	12.5	7.6	12.4	7.0	9,991	15,557	40.3	12.6	17,094	2,653
Provo - 363 N. University Ave. (49-049-0005)	1	12.5	7.6	12.4	7.0	9,991	15,557	70.5	12.6	17,094	2,653

[a]Monitor area defined by census tracts immediately surrounding monitor site (except Birmingham monitor area defined by block group).
[b]Region defined by Metropolitan Statistical Area (MSA) except for Fairbanks North Star Borough, Los Angeles PMSA, Imperial County (Calexico), and Flathead County (Kalispell).
[c]Share of workers 16 years and older that do not drive alone or carpool to work.
[d]County of Los Angeles used for Reseda monitor site.
[e]Special monitor located to monitor industrial site; monitor area defined by block group.
Sources: See Table 1-1 and U.S. Census Bureau 2000a.

the regional average, is an anomaly and reflects a recent influx of affluent residents to downtown Denver—residents choosing to live in a high-density, high-traffic area.

In addition, the number of employed residents who do not drive to work is higher than for the region as a whole in all but three of the monitor areas and is four or more times higher in monitor areas in Spokane, Washington, Denver, Colorado, and Provo, Utah. Residents who walk, bike, or ride transit are likely to spend more time within the monitor area during their commutes and may experience greater exposure to high CO levels.

To a limited degree, these demographics may explain the high levels of CO recorded in many monitor areas. Although the residents of these areas are less likely to drive to work, they are more likely to own older vehicles, which in turn are more likely to be high-emitters (Rajan 1993; Granell 2002). For example, Singer and Harley (1996, 2000) observed a much higher fraction of older vehicles near the Lynwood monitor. The vehicles observed in their study included vehicles passing through the area as well as vehicles owned by local residents. The high emissions rates for older vehicles may offset lower total amounts of driving. In addition, the relatively high population densities in all but one monitor area (Table 1-7) suggest higher concentrations of traffic in these areas and thus higher concentrations of pollutants. However, the traffic generated locally is likely to represent a small fraction of the total traffic in and around most of these monitor areas.

These preceding demographics suggest the need for continued attention to the CO problem from the standpoint of environmental justice. In 1994, President Clinton signed Executive Order 12898, Federal Actions to Address Environmental Justice in Minority Populations and Low-Income Populations. The order was related to Title VI of the Civil Rights Act of 1964 and required federal agencies "to achieve environmental justice by identifying and addressing disproportionately high and adverse human health and environmental effects, including the interrelated social and economic effects of their programs, policies, and activities on minority populations and low-income populations in the United States." The order also stipulated that in reviewing other agencies' proposed actions under Section 309 of the CAA, "EPA must ensure that the agencies have fully analyzed environmental effects on minority communities and low-income communities, including human health, social, and economic effects " (EPA 1998a).

EPA and the FHWA have issued their own interpretations of the environmental justice requirement. EPA defines environmental justice as "the fair treatment of people of all races, cultures, and incomes with respect to the development, implementation, and enforcement of environmental laws and policies, and their meaningful involvement in the decision making processes of government" (emphasis in original). According to EPA, fair treatment requires that EPA conduct its "programs, policies, and activities that substantially affect human health and the environment in a manner that ensures the fair treatment of all people, including minority populations and/or low-income populations" and that EPA ensure "equal enforcement of protective environmental laws for all people, including minority populations and/or low-income populations" (EPA 2001d). As interpreted by the FHWA, environmental justice includes not just minimizing adverse effects but also the preventing of "the denial of, reduction in, or significant delay in the receipt of benefits by minority and low-income populations" (FHWA 1998). This requirement should apply to benefits from federal policy such as improvements in air quality.

Both of these interpretations suggest that the remaining locations experiencing high CO concentrations represent a potential environmental justice concern. According to EPA's final guidance for incorporating environmental justice concerns into National Environmental Protection Act (NEPA) compliance analyses, a minority population is present "if the minority population percentage of the affected area is 'meaningfully greater' than the minority population percentage in the general population or other 'appropriate unit of geographic analysis'" (EPA 1998a). This is clearly the case for areas surrounding most of the monitors registering CO exceedances since 1995. In addition, EPA notes the following:

> Minority communities and low-income communities are likely to be dependent upon their surrounding environment (e.g., subsistence living), more susceptible to pollution and environmental degradation (e.g., reduced access to health care), and are often less mobile or transient than other populations (e.g., unable to relocate to avoid potential impacts). Each of these factors can contribute to minority and/or low-income communities bearing disproportionately high and adverse effects (EPA 1998a, p. 57).

The extent of the areas with CO concentrations that exceed the NAAQS is unclear because the number of monitors in each area is limited. Therefore, measures of the exposures to CO experienced by low-income and minority populations are imperfect. While the declining number of exceedances of the CO standard and the design of the standard to ensure a

> **Recommendations: Spatial Distribution of CO**
>
> EPA should employ air quality modeling and saturation studies in CO problem areas to better characterize the spatial distribution of CO and the populations affected. The information garnered can be used to improve site selection fro permanent monitoring, to improve model performance, and to address possible environmental equity issues. Programs targeted to local conditions can be developed using this information. These results should also be linked to health impact studies in these locations. In particular, EPA should try to better understand the upper end (higher CO levels) of the distribution of ambient exposures to motor-vehicle emissions that occur in most CO hot spots.

reasonable margin of safety are encouraging, the correlation between CO and other pollutants creates uncertainty in the degree to which "adverse human health and environmental effects" might be occurring in these low-income and minority communities. Nevertheless, the demographic patterns suggest that the impacts that occur are disproportionately high in these areas. In addition to providing an important impetus for the continuation of efforts to eliminate CO exceedances, these results suggest a need for monitoring and personal exposure research programs designed to more fully characterize the distribution of CO and other mobile-source-related pollutants.

2

Contributions of Topography, Meteorology, and Human Activity to Carbon Monoxide Concentrations

INTRODUCTION

Topography, meteorology, and human activity contribute to high carbon monoxide (CO) concentrations in some areas that exceed the National Ambient Air Quality Standards (NAAQS). Despite the decline in national ambient CO concentrations, maintaining the 8-hour standard of 9 parts per million (ppm) has been a particular challenge for some locations (see Table 1-1). Even when attainment of the standard has been achieved, there remains a vulnerability to future exceedances. Expressed mathematically, there is a nonzero probability of nonattainment in a future year.

After a more detailed discussion of topography and meteorology, this chapter will discuss the seasonal, weekly, and diurnal patterns in CO concentrations measured in some CO problem areas. These patterns help describe some of the physical and human factors contributing to the CO problem in these locations. The chapter discusses vulnerability to future exceedances, including a brief description of statistical approaches. The chapter concludes with illustrative examples of the factors contributing to the CO problems in Calexico, California; Lynwood, California; Fairbanks, Alaska; Las Vegas, Nevada; and Denver, Colorado.

The committee identified four factors that contribute to the difficulties that the cities listed in Table 1-1 have had in meeting the NAAQS for CO:

1. *Unfavorable topography.* Low lying areas surrounded by higher elevations on three or more sides are vulnerable to CO buildup.
2. *Unfavorable meteorology.* Stagnant winter conditions characterized by ground level temperature inversions (see definition below) and low windspeeds inhibit vertical mixing of CO.
3. *Significant local CO emissions.*
4. *High concentrations of CO transported from nearby areas.*

Higher elevations with lower air densities tend to have higher CO concentrations (in ppm) for a given emission flux.[1] Denver's average air density is 85% of that at sea level. Lower oxygen density can increase CO emissions rates in older vehicles. Topography also can affect meteorological conditions in a variety of ways, as described below.

Meteorology can influence pollutant concentrations through its effects on atmospheric mixing height, windspeeds and wind direction, and atmospheric water content (humidity). Humidity is a factor because dry climates and higher elevations tend to have lower total water columns overhead. Because water vapor is an important greenhouse gas (infrared radiation from the earth's surface is absorbed by water molecules and reradiated back down, warming the surface), reduced water vapor allows infrared radiation to pass into space, producing ground level temperature inversions after sundown[2] and lower mixing heights. These lower mixing heights, combined with high evening traffic emissions, can lead to pollution buildup near the ground. It is noteworthy that all of the cities listed in Table 1-1, except Birmingham, are west of the Mississippi River, where the air tends to be drier. Two are in Alaska, where high latitudes and low winter temperatures result in reduced solar heating at midday and atmospheric conditions are typically dry.

---

[1] Considering CO emissions as an area source, the emission flux is the mass of CO produced per square kilometer per hour.
[2] This radiation into space is the reason temperatures drop so rapidly at night in the desert and in the mountains.

METEOROLOGY AND TOPOGRAPHY

Background

The ease with which air can mix vertically to disperse pollutants depends critically on how the air temperature changes with altitude. Warmer air is less dense (more buoyant) and tends to rise[3] and cool as the pressure decreases and volume expands. If the vertical temperature profile decreases by 1°C/100 m (the adiabatic lapse rate) or more, the air mixes freely as warmer air from below moves upward. If the temperature decreases more slowly than 1°C/100 m, or increases with altitude (called an inversion),[4] vertical mixing is inhibited. The faster the air temperature increases with altitude during an inversion, the more strongly mixing is resisted.

Inversion Types

There are several atmospheric processes that can form inversions, as illustrated in Figures 2-1 and 2-2. Cooling of the air near the ground as a result of infrared radiation into space after sunset can create a surface-based inversion, like that shown in Figure 2-1(a), and can produce a thermodynamically stable layer, which tends to trap pollution near the ground. Horizontal advection of warm air creates a high-altitude inversion and can similarly increase the stable temperature stratification aloft (Figure 2-1[b]). Figure 2-1(c) shows the situation with both surface-based and high altitude inversions. In a subsidence inversion (Figure 2-2), a surface-based inversion can be strengthened by warm air that sinks and is warmed further as a result of compression. Each of the inversion types reduces the atmosphere's ability to mix through the inversion level, allowing pollution generated below the inversion level to accumulate.

---

[3]This is the principle on which hot air balloons operate. Vehicle exhaust from a tailpipe is typically 50-100°C warmer than the surrounding air; however, it is rapidly diluted and cooled as it is mixed into the ambient air.

[4]An "inversion" in meteorology is defined as "a departure from the usual decrease or increase with altitude of the value of an atmospheric property" (Geer 1996). The term is generally used to refer to a situation where temperature shows an increase with altitude rather than the usual decrease.

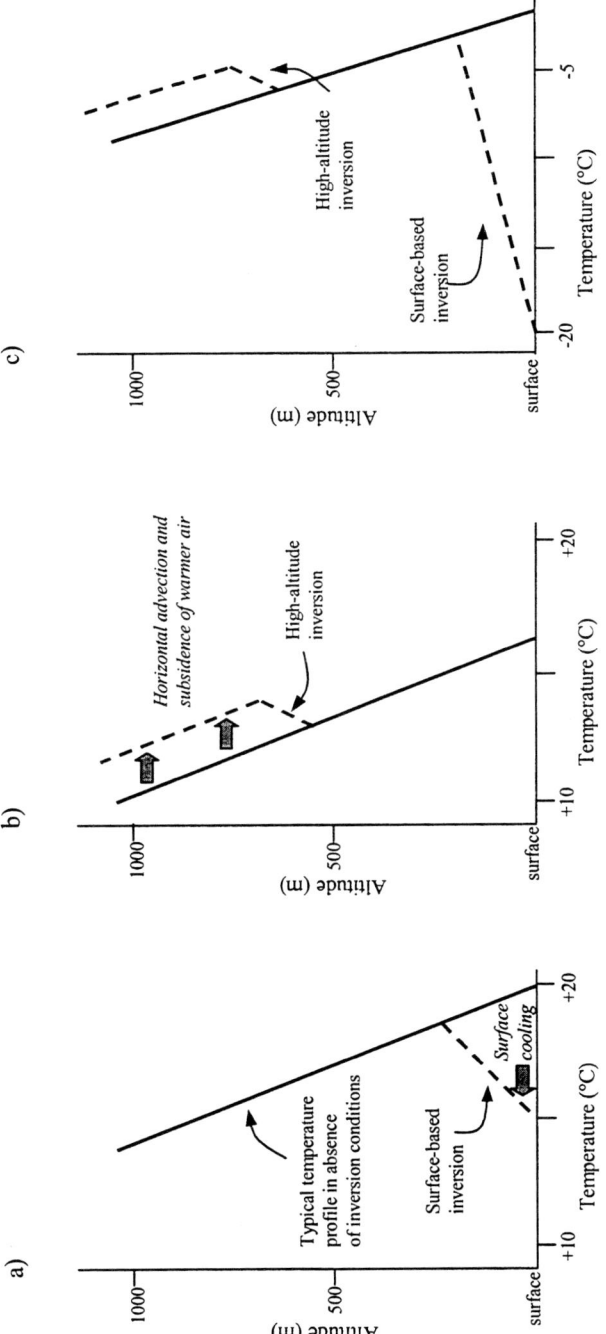

FIGURE 2-1 Schematics of (a) surface-based inversions, (b) high-altitude inversions, and (c) typical Alaskan temperature profile with both surface-based and high-altitude inversions. Solid lines indicate the temperature profile in the absence of inversions. Dashed lines indicate temperature profiles affected by inversions.

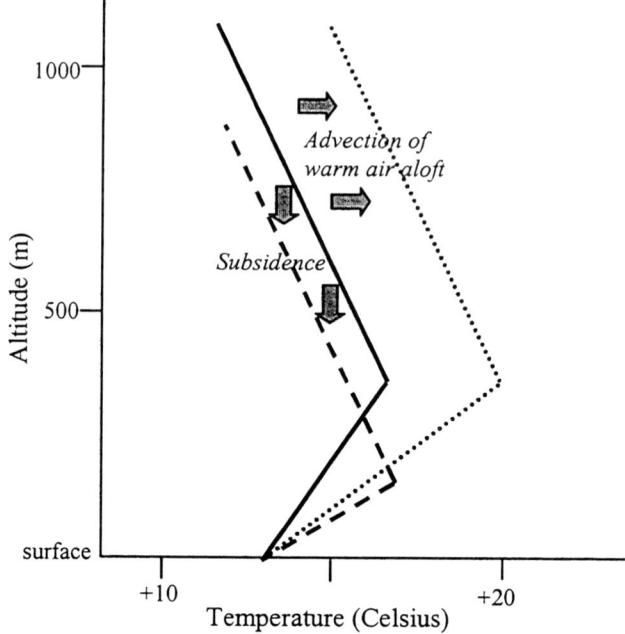

FIGURE 2-2 Schematic of how an existing surface-based inversion (solid line) can be strengthened by subsidence (dashed line) or by advection of warm air aloft (dotted line).

Recirculation

Atmospheric flow eddies can recirculate the air one or more times. When pollution is emitted into these circulations, pollutant concentrations can increase over time. Figure 2-3 illustrates such a recirculation pattern in a trapping valley.

Sea and land breezes represent an additional cause of atmospheric recirculation (Segal and Pielke 1981). As shown in the modeling study of Eastman et al. (1995), at least 70% of pollution recirculates with the summer Lake Michigan sea breeze. These results mirror the observations of Lyons et al. (1995). That study shows that Gaussian-type models fail to replicate the recirculation and the complex dispersion patterns that result when spatial variations in sensible heat fluxes exist at the surface (Pielke and Uliasz 1993).

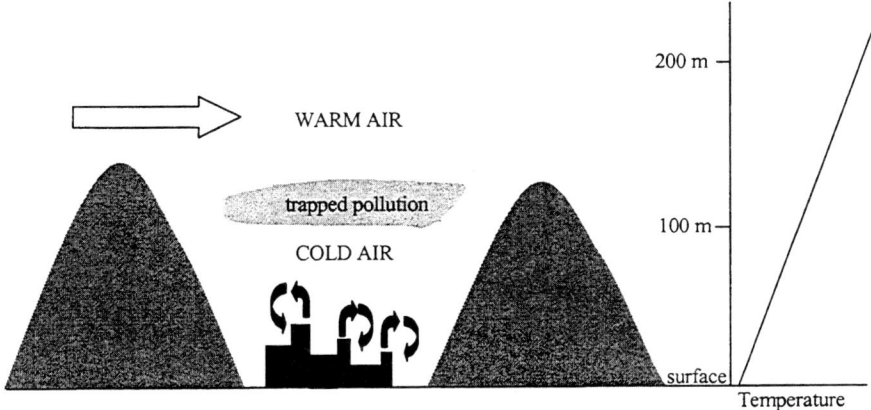

FIGURE 2-3 Schematic of a trapping valley. The temperature profile in the valley is shown on the right.

Stagnation

When air does not move significantly over tens of hours or more, the atmosphere is said to be stagnant. Stagnation can occur because of weak winds and/or the trapping of air (see Figure 2-3). When the atmosphere is stagnant, emitted pollution can accumulate over time. Simple box models, such as those discussed in Pielke et al. (1991) and in Appendix C, can be used to estimate pollution buildup associated with stagnation. Figure 2-3 illustrates air stagnation in a trapping valley.

Influence of Topography on Meteorological Conditions

Pielke (2002) discusses the influence of terrain on atmospheric conditions under strong and weak large-scale winds. The following discussion illustrates that the direction of movement of air is actually quite complicated in complex terrain. In contrast to flat terrain, the wind flow in valleys can go in almost any direction, depending on the relative importance of the different forcing mechanisms. These mechanisms will affect the dispersion of CO. Under strong flow, for instance, large upward and downward motions are produced, which can enhance pollution dispersion. Under weak flow, mountain-valley winds that are a result of diurnal warming and cooling of the terrain surface can occur (Figure 2-4). These local winds transport and disperse pollution, although recirculation can also occur. Mountain-valley winds can occur on relatively small scales (in which case they are called upslope and drainage flows) or on larger scales (where broad ascent and descent patterns occur). The resulting wind flows can be quite complex.

Figure 2-5 illustrates the differences in the diurnal variation of the valley wind direction as a function of the wind direction above the valley (the geostrophic wind) for four distinct mechanisms that can control the wind direction. According to Pielke (2002), these four physical mechanisms operate as follows: (1) thermally driven winds are independent of the above-valley winds and are controlled by locally developed valley pressure gradients; (2) downward momentum transport of the above valley winds (as is associated with a deep convective boundary layer) produces similar wind directions at all levels; (3) forced channeling occurs when the valley flow alignment is dependent on whether the above-valley flow has a net flow down- or up-valley; and (4) pressure-driven channeling (which is out of phase with forced channeling) occurs when the winds in the valley respond only to large-scale horizontal pressure, not to the winds that occur above the valley. Without terrain, the airflow would be nearly parallel to the isobars. With other local flows involved (such as sea and land breezes), wind flow is more complex (Pielke 2002).

Large-Scale Meteorological and Climatological Events and Their Impact on Attainment

Local air quality can be affected by large-scale meteorological and climatological events. CO exceedances may have patterns that are related to the occurrence of synoptic-scale meteorological events or climatological events, such as the El Niño Southern Oscillation (ENSO). Changes in the frequency of large-scale events could affect a location's ability to come into and maintain compliance with the NAAQS for CO. The committee explored the potential effects of large-scale meteorological and climatological phenomena on local CO episodes in three cities: Lynwood, California; Fairbanks, Alaska; and Denver, Colorado. It should be noted, however, that the impact of climate and meteorological variability on air quality, including CO and related pollutants, is an area requiring more research.

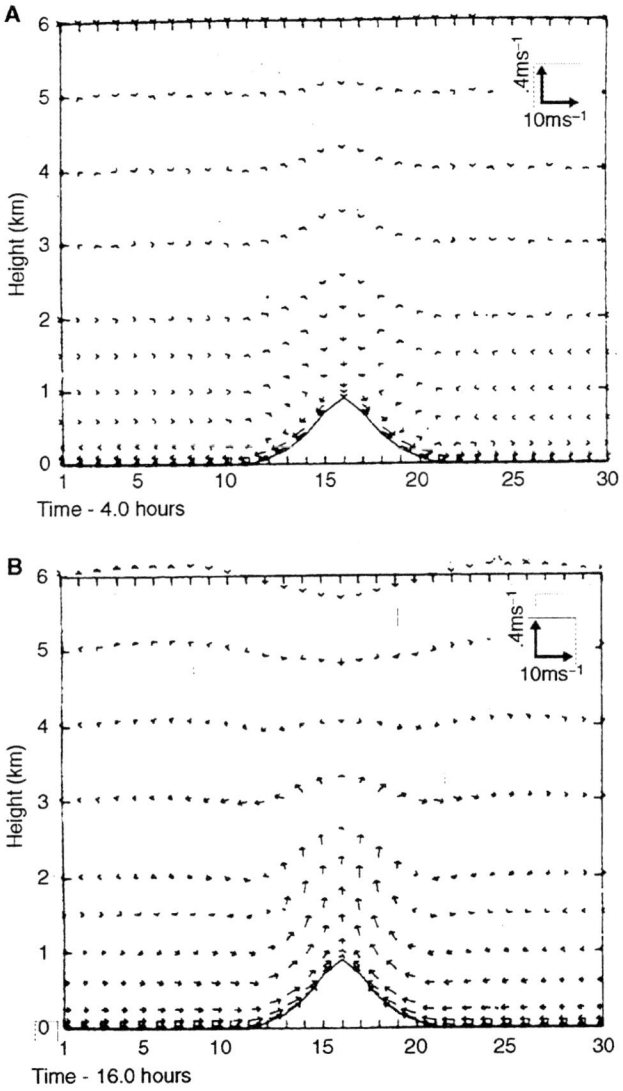

FIGURE 2-4 Two dimensional simulation of (a) nocturnal drainage flow and (b) upslope flow with no prevailing synoptic flow, with an input condition typical of summer in midlatitudes. Source: Mahrer and Pielke 1977. Reprinted with permission; copyright 1977, American Meteorological Society.

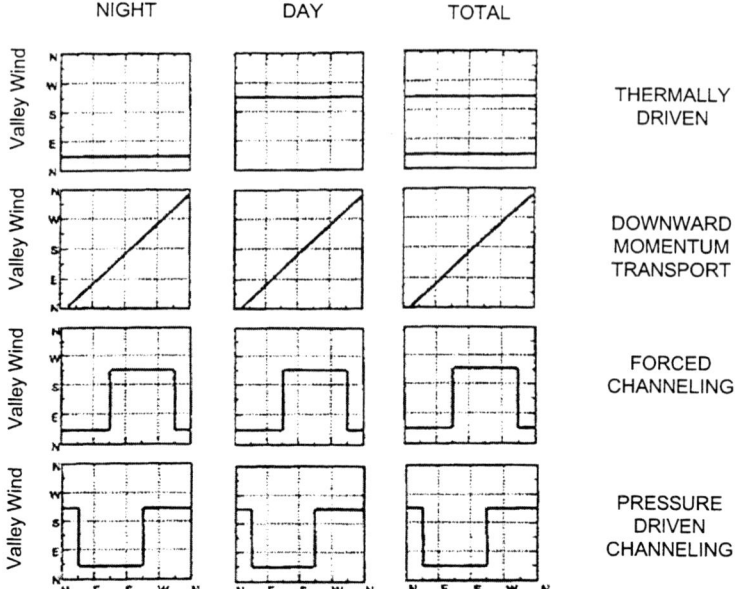

FIGURE 2-5 Relationships between above-valley (geostrophic) and valley wind directions for four possible forcing mechanisms: thermal forcing, downward momentum transport, forced channeling, and pressure-driven channeling. The valley is assumed to run from northeast to southwest. Source: Whiteman and Doran 1993. Reprinted with permission; copyright 1993, American Meteorological Society.

Lynwood, California

Lynwood's local air quality may be influenced by ENSO. Historically, El Niño recurs every 3-7 years when sea-surface temperatures (SSTs) in the equatorial Pacific Ocean off the South American coast become warmer than normal. La Niña is essentially the opposite of El Niño, and exists when cooler-than-usual ocean temperatures occur near the equator between South America and the International Date Line.

During an El Niño, the months of October through March tend to be wetter than usual in a swath extending from southern California eastward across Arizona, southern Nevada, Utah, and New Mexico, and into Texas. Almost all of the major flood episodes on main rivers in southern California have occurred during El Niño winters. During La Niña years, dry conditions are produced on the equator in the Pacific Ocean. La Niña generally does not affect the United States as much as El Niño; however, strong La Niñas have been linked to dry seasons in southern California. In Lynwood, CO exceedances may tend to increase during strong La Niña years when dry and stable atmospheric conditions are produced: Conversely, CO exceedances may tend to decrease during El Niño years. Although some studies have explored the effects of El Niño on ozone levels (e.g., Chandra et al. 1998), as of yet no studies have examined correlations between ENSO and CO exceedances. Further research may be needed in this area, including an assessment of how ENSO affects conditions that control concentrations of the pollutants associated with CO (i.e., air toxics and PM).

Fairbanks, Alaska

In Fairbanks, Alaska, all exceedances of the 8-hour CO standard from 1996 through 2001 occurred when a low-pressure system in or near the Gulf of Alaska produced southeasterly geostrophic winds. These winds, which travel over the Alaska Range, are associated with the counterclockwise geostrophic flow around the low-pressure system. One hypothesis for the coincidence of CO exceedances with this synoptic-scale meteorological event is that the warm-air advection aloft reinforces the radiative ground-level inversion. The downward movement of air over Fairbanks also exerts a stabilizing influence on inversions. It is not known, however, what fraction of nonexceedance days has such meteorological conditions or how many of the exceedance days before 1996 had these conditions. Nonetheless, the surface pressure gradient observed during all six exceedances from 1996 to 2001 must have some significance. However, further research over a longer period of time is needed to better understand the relationship.

Denver, Colorado

In the past, CO exceedances in Denver, Colorado, have coincided with the occurrence of lee troughs—lines of surface low pressure on the lee side (the side that is sheltered from the wind) of a mountain range. The air coming over the mountains sinks and warms and, at the same time, the lowering of pressure at the surface along the foothills draws colder air from the plains and lowlands areas back towards the mountains. Thus, the air between the surface and 100-300 m becomes colder as the air above becomes warmer, enhancing the inversion. Neff and King (1991) characterized lee trough history for the 1980s and early 1990s. Lee troughs often occur several times each week during the winter months and are a key precursor to high CO levels in Denver. However, the effect of lee troughs on CO exceedances has not been studied since the mid-1990s because of the decline in exceedances. The decline in CO exceedances is mainly due to lower vehicle emissions, but Neff (2001) noted that there was also a decline in the occurrence of lee troughs during the late 1980s, which perhaps reduced the frequency and the severity of conditions producing CO exceedances. Future studies also should explore the association of lee troughs, the Arctic Oscillation, and air quality. When cold-air arctic outbreaks occur, they usually provide a snow cover, which strengthens the ground-level inversion. Fewer arctic outbreaks over the Great Plains could help decrease the potential for pollution episodes in Denver. Denver has had a decade-long period without the long-term snow cover and associated light winds that tend to promote atmospheric stagnation, which can lead to CO buildup. The lack of conditions conducive to high CO means less susceptibility to CO exceedances.

TEMPORAL PATTERNS OF CO CONCENTRATIONS

CO concentrations show seasonal, weekly, and diurnal patterns reflecting the temporal patterns in emissions and meteorology. Figures 1-1 and 1-2 show seasonal patterns in the numbers of days with exceedances of the 8-hour CO standard. Figures 2-6 and 2-7 also show patterns in the total numbers of exceedance days by month for Lynwood, California, and Fairbanks, Alaska, for periods of approximately 30 years. Lynwood exhibits a very symmetrical pattern, with the maximum number of exceedance days in December, when the winter solstice (shortest day, least solar radiation) occurs. Fairbanks exhibits the maximum number of exceedances in January. The considerably greater numbers of exceedance days in Fairbanks in January compared with November, and in February compared with October, are attributed to reduced cloud cover in the winter months compared with the autumn months.[5] Clear skies in January and February contribute

---

[5]See Figure 2-3 in the interim report (NRC 2002) for Fairbanks cloud cover by month.

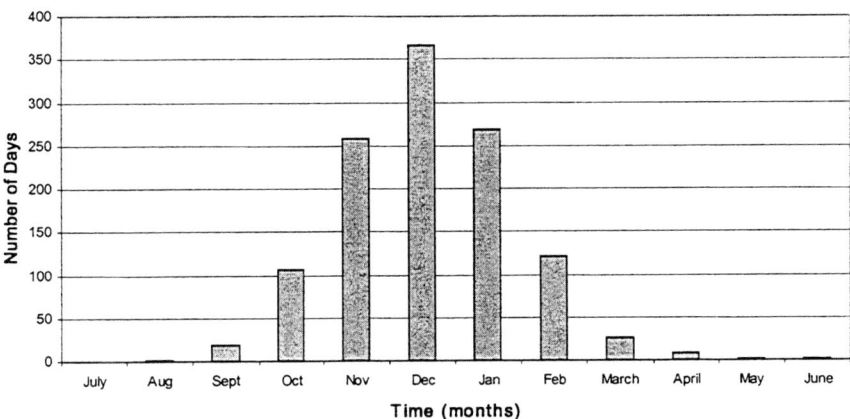

FIGURE 2-6 Number of days with exceedances of the 8-hour CO standard by month in Lynwood, California, between July 1973 and June 2000. (Data were missing for December 1997.)

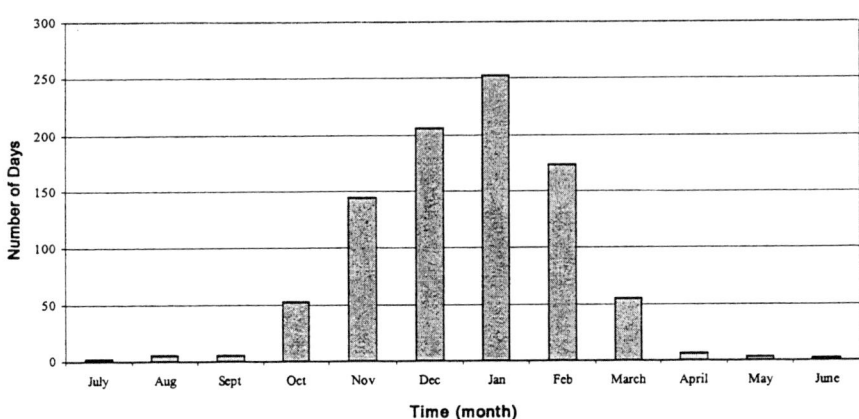

FIGURE 2-7 Number of days with exceedances of the 8-hour CO standard by month in Fairbanks, Alaska, between July 1972 and June 2001. Numbers include days on which exceedances occured at any of the three monitoring sites.

84   Managing CO in Meteorological and Topographical Problem Areas

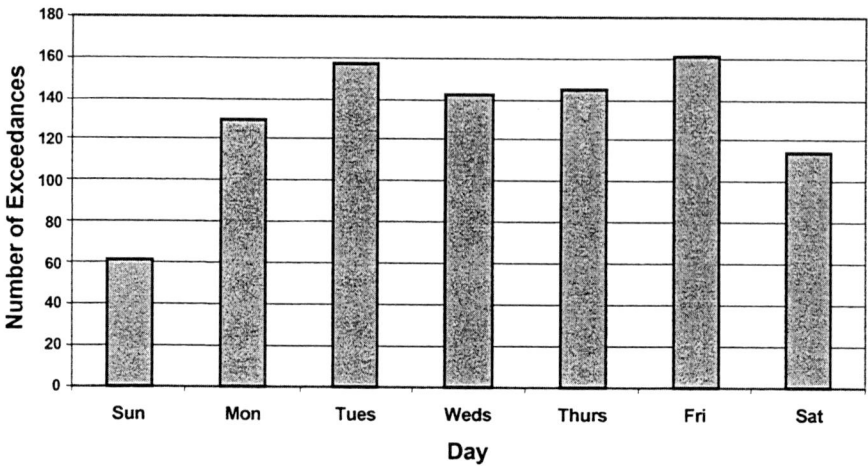

FIGURE 2-8 Number of days with exceedances of the 8-hour CO standard by day of the week in Fairbanks, Alaska, over the period July 1972 through June 2001. The numbers include days on which exceedances occurred at any of the three monitoring sites. Source: Data provided by Paul Rossow, Fairbanks North Star Borough.

to stronger ground-level inversions because of increased radiative heat loss. Weekly patterns of CO concentrations and exceedance days occur because of patterns in vehicle use. Higher emissions typically are recorded on weekdays rather than weekends. Vehicle use is especially low on Sunday. This can be seen in Figure 2-8 (above) for Fairbanks. The diurnal patterns of CO concentrations in the cities with the most severe CO problems are discussed below.

Comparison of Diurnal Variations in CO Concentrations in Ten Cities

Insights into the fundamental physical and human processes involved in the buildup of CO in the atmosphere and how these processes vary among areas can be gained by comparing the diurnal variations of CO concentrations in the cities that have had the most difficulty meeting the 8-hour CO NAAQS (Table 1-1), excluding Birmingham, Alabama. The general diurnal variation of CO is described in the most recent criteria document for CO (EPA 2000a). There are two maxima, a stronger one at 7:00-9:00 a.m. and a weaker one at 6:00 p.m. (hour 18), and two minima, a deeper one at 5:00 a.m. and a shallower one at 2:00-3:00 p.m. This general pattern can be compared with the diurnal variations measured at the monitors listed in Table 1-1. Figure 2-9 shows the diurnal variations of mean and maximum 1-hour average CO concentrations[6] for nonholiday weekdays during the winter of 1999-2000 (the most recent winter for which hourly data were available for the 10 cities) shown in order of decreasing latitude (north to south).

The mean concentrations (broad lines in Figure 2-9) show some interesting patterns, although these patterns may be due in part to local traffic flow. Although most of the monitors showed both morning and evening maxima, the one at the Post Office building in Fairbanks showed only one maximum—in the evening—a factor noted in the criteria document (EPA 2000a). A number of the more northern sites show indistinct afternoon minima. The very deep minima in Las Vegas and Calexico in the early afternoon suggest that solar heating at these southern latitudes is able to disperse the CO effectively at that time of day. (To verify that low afternoon CO concentrations are due to meteorology rather than to very low afternoon emissions would require an analysis of traffic flow near the monitors.)

Fairbanks, Spokane, the Anchorage 1 site (AIRS ID 02-020-0037), Kalispell, and Denver each show a daily maximum at 6:00 p.m., attributable to local evening rush-hour traffic.[7] The times of evening maxima in other areas varied widely; those in Lynwood, Calexico, and Phoenix were at 10:00-11:00 p.m.[8] Las Vegas showed an unusually broad evening maxi-

---

[6]For each hour of the day (starting at midnight), the mean of 1-hour average CO concentrations for the 85 nonholiday weekdays was determined, along with the standard deviation and maximum for that hour. Weekend days and major holidays (Thanksgiving, Christmas, New Year's Eve, and New Year's Day) were not included.

[7]Because the CO concentration may take time to build up, the maximum in concentration might be expected to lag the maximum in traffic flow somewhat. Modeling shows that the lag time depends on the windspeed but is typically an hour or less.

[8]Times of these maxima are much later than when people are likely to be out exercising.

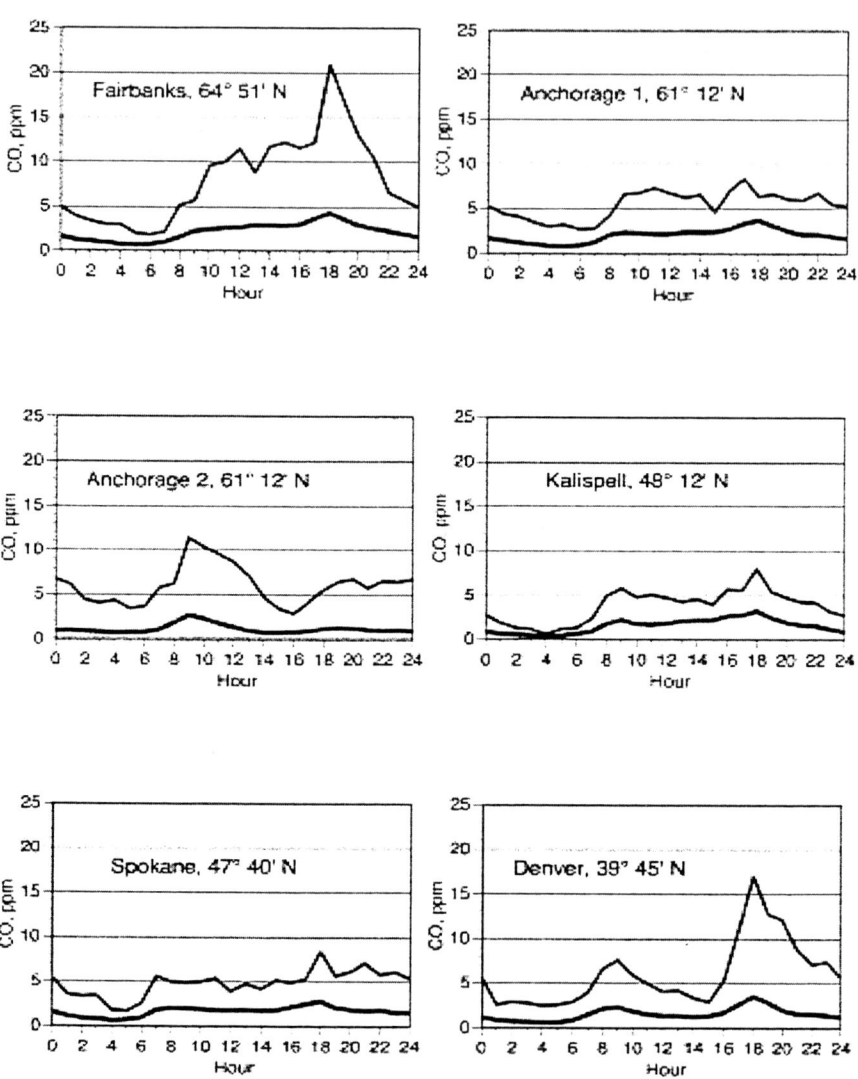

FIGURE 2-9 Diurnal variations of mean (heavy line) and maximum (light line) 1-hour CO concentrations at monitoring sites listed in Table 1-1 for 85 nonholiday weekdays during the winter (November through February) of 1999-2000. Sites are shown in order of decreasing latitude (north to south). The

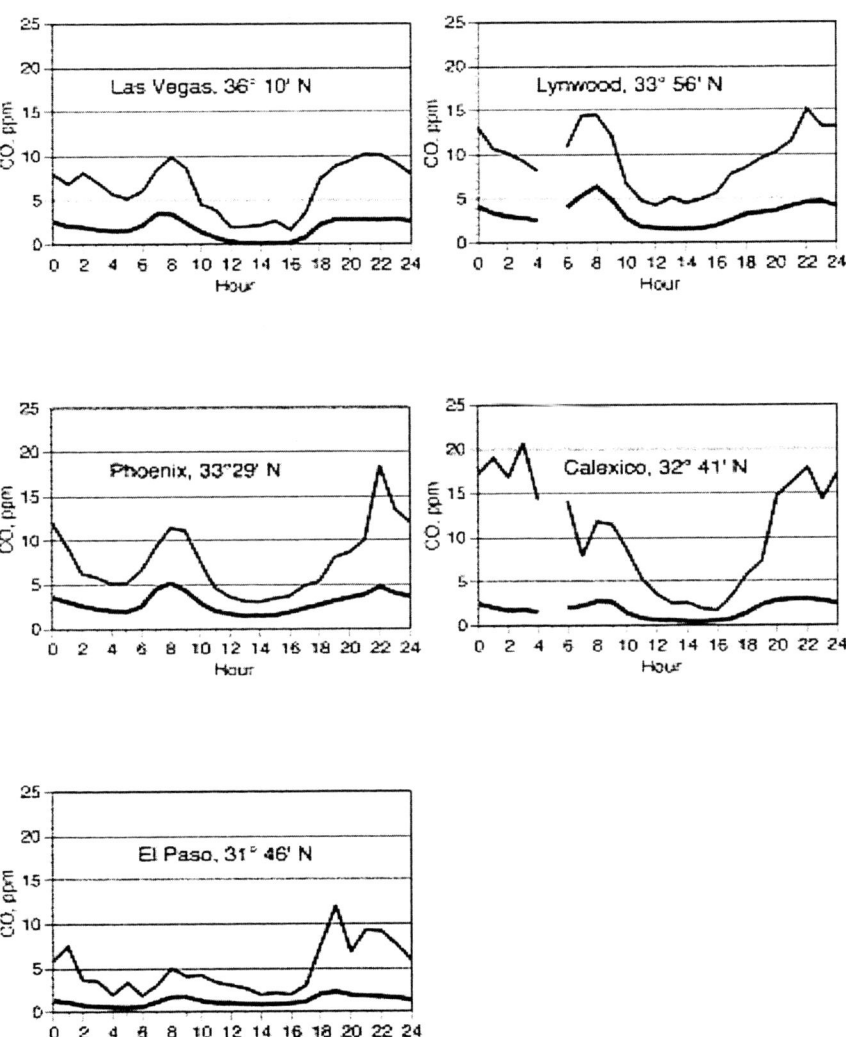

horizontal scale refers to the end of the 1-hour averaging period. There are breaks in the curves for Lynwood and Calexico because the period from 4:00 a.m. to 5:00 a.m. was used for instrument calibration.

mum that was essentially flat from 7:00 p.m to 12:00 p.m., possibly due to late-evening local traffic.

The Anchorage 1 and Anchorage 2 sites, located downtown and in a residential neighborhood, respectively, differ in that the latter exhibits a morning maximum. This is likely due to cold-start emissions and periods of engine idle before motorists leave home for work. The average warm-up idle for morning commuters in Anchorage under these very cold conditions is approximately 10 min, some idling as long as 30 min (Morris and Taylor 1998).

The highest 1-hour CO concentrations observed in Lynwood, Phoenix, and Calexico (the light lines in Figure 2-9) were unusual. They occurred at 10:00 p.m. or 3:00 a.m., and were consistent with CO transport into these areas from outside sources. Calexico is adjacent to Mexicali, a much larger Mexican city to the south, and Lynwood is surrounded by metropolitan Los Angeles. (See the illustrative examples of Calexico and Lynwood later in this chapter.)

The time of day when an area experiences its daily maximum CO concentration depends on a variety of factors, including the time dependent diurnal emission rate in the area (including vehicle cold starts and traffic), meteorological variables (especially windspeed and inversion strength near the ground), and in some cases transport in from surrounding areas. Simulations of the six most recent exceedance days in Fairbanks (five of which occurred in February) showed that vehicles leaving downtown for home combined with strong evening temperature inversions at sundown yielded the highest CO concentrations at 5:00-6:00 p.m. (NRC 2002).

VULNERABILITY TO FUTURE EXCEEDANCES

Once a nonattainment area is able to reduce its ambient CO concentrations and has eight clean quarters (no violations of the 8-hour CO standard over 2 years), it can apply for a change in attainment status. A state implementation plan (SIP) is required and must show that maximum 8-hour average CO concentrations can be expected to remain below 9 ppm for the next 10 years. The confidence of the projection depends not only on expected future emissions (for example, on the basis of expected vehicle-miles traveled (VMT) and fleet composition) but also on the variability of CO concentrations for a given annual emissions inventory. Some cities show more variability than others.

Effects of Meteorology and Emissions on Vulnerability to Future Exceedances

This section examines the effects that meteorology and emissions have on the vulnerability of locations to future CO exceedances. Table 2-1 summarizes the characteristics of the mean diurnal patterns shown in Figure 2-9, indicating the time of the daily maximum at each site for the average nonholiday weekday, the mean CO concentration at that time, and the standard deviation. The next-to-last column provides the coefficient of variation (COV)—the ratio of the standard deviation to the mean—for the daily maximum; the largest COVs (>0.7) are shown in italics. The last column gives the average CO concentrations for the 1999-2000 winter season, including weekends and holidays.

The variability indicated in the COV column in Table 2-1 is the result of variability in both emissions rates and meteorological factors. Traffic volumes at 6:00 p.m. (hour 18) on significant roadways in Fairbanks, Alaska, during nonholiday winter weekdays show little variation, with a COV for traffic of only 0.08 ± 0.01.[9] Much greater variability is exhibited in CO concentrations than in traffic. Figure 2-10 compares the variability in daily average traffic on Cushman Avenue in Fairbanks with the variability in daily (24-hour) average CO concentrations measured at the Post Office monitor on the same road during the winter of 1999-2000. In each case, the data are sorted into two categories: (1) nonholiday weekdays, and (2) weekends and major holidays. Then the data are sorted by decreasing daily average value within each of those categories. The mean daily average traffic for winter weekdays in Figure 2-10 was 616 vehicles per hour, with a standard deviation of 55 (COV = 9.0%), and the mean daily weekday average CO concentration was 2.2 ppm, with a standard deviation of 1.1 (COV = 50%).[10] These figures confirm that CO concentrations were considerably more variable than traffic flows. The highest and second-highest daily average CO values for that winter occurred on Tuesday, February 8, 2000, and on Monday, November 19, 1999. Both were exceedance days,

---

[9] This mean (0.08) and standard deviation (0.01) are based on an analysis of traffic on three roads in downtown Fairbanks during winters of 1995-1996 through 2000-2001. For the locations of counters and monitors see Figure 2-4 in the interim report (NRC 2002).

[10] This ratio of 0.50 differs from the 0.81 that appears in Table 2-1 because the former refers to a 24-hour mean and the latter to a 1-hour mean.

TABLE 2-1 Summary of Diurnal Average CO Behavior for Nonholiday Weekdays During Winter, 1999-2000

City and State	AIRS ID	Daily Maximum[a]				Winter Mean[c]
		Hour	Mean[b]	Standard Deviation[b]	COV[b]	
Calexico, CA	06-025-0005	22	3	3.4	*1.11*	1.8
Lynwood, CA	06-037-1301	8	6.4	3.5	0.54	3.3
Fairbanks, AK	02-090-0002	18	4.3	3.6	*0.81*	2.2
Phoenix, AZ	04-013-0022	8	5.2	2.3	0.44	2.8
Spokane, WA	53-063-0040	18	2.8	1.5	0.52	1.7
Las Vegas, NV	32-003-0561	7	3.5	2.3	0.66	1.8
Anchorage, AK (site 1)	02-020-0037	18	3.7	1.3	0.35	2
Anchorage, AK (site 2)	02-020-0048	9	2.7	2.9	*1.06*	1.2
El Paso, TX	48-141-0027	19	2.3	2.1	*0.92*	1.3
Kalispell, MT	30-029-0045	18	3.3	1.4	0.44	1.6
Denver, CO	08-031-0002	18	3.5	2.8	*0.78*	1.6

[a]Hour refers to a 24-hour clock. CO concentrations are in parts per million by volume.
[b]The mean, standard deviation, and COV (the ratio of the standard deviation to the mean) at each site are for 85 nonholiday weekdays at the hour shown for the daily maximum; COVs >0.7 are shown in italics.
[c]The winter mean at each site is for all hours for which there are data, from the beginning of November 1999 to the end of February 2000, including weekends and holidays.
Abbreviations: AIRS ID, aerometric information retrieval system identification number; COV, coefficient of variation.

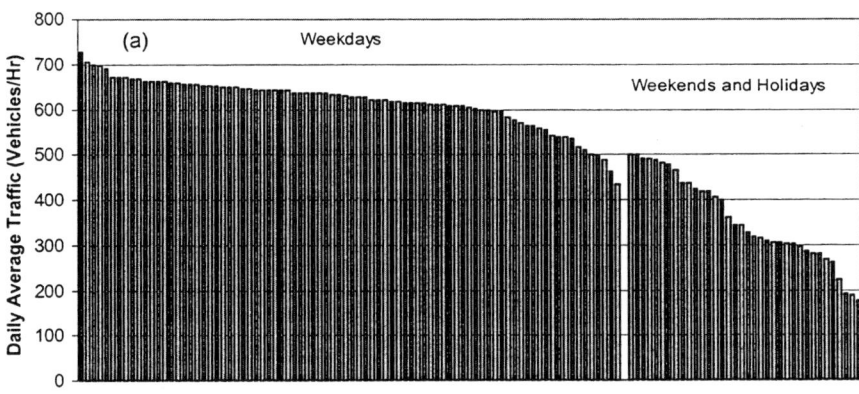

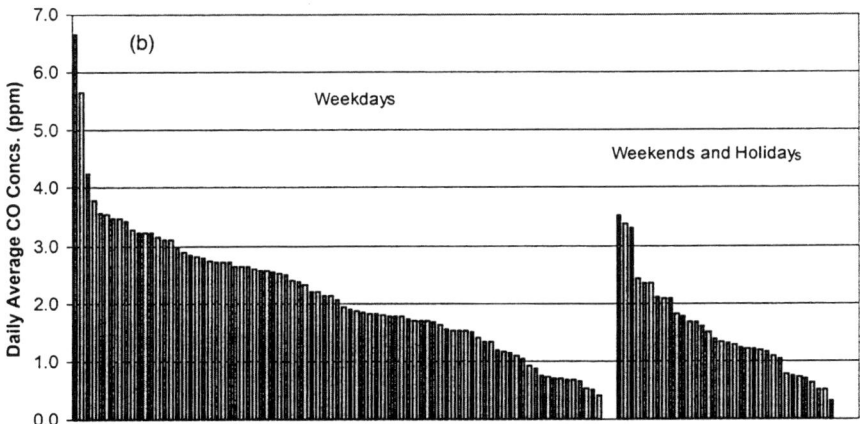

FIGURE 2-10 Daily average (a) traffic on Cushman Avenue and (b) CO concentrations at the Post Office on Cushman in Fairbanks during the winter of 1999-2000. Values for weekdays are rank ordered, as are those for weekends and major holidays (Thanksgiving, Christmas, New Year's Eve, and New Year's Days). Days with data missing are indicated by missing bars. The daily average traffic was obtained by diving the total traffic count each day by 24 hours.

with maximum 8-hour average CO values of 11.5 ppm and 11.2 ppm, respectively. Figure 2-11 shows a scatter plot of the data used for Figure 2-10. The two exceedance days (the top two points) had traffic flows (600-

700 vehicles per hour) that produced many much lower average CO concentrations.

An exceedance that occurred in Fairbanks on Saturday, January 11, 1997, highlights the importance of meteorological factors. The maximum 8-hour average CO concentration that day was 13.3 ppm even though the daily mean traffic on Cushman Avenue was only 437 vehicles per hour. The daily average lower inversion strength (measured between 3 and 10 m above the ground) was 18.6°C/100 m, the average windspeed was 0.8 MPH, and the average temperature 5.5°F.[11] Although traffic is not the only factor determining CO emissions rates (cold-start and idling emissions are also important in Fairbanks), such an exceedance indicates that meteorology may be able to produce exceedances despite emissions reductions.

Two cities with the same average CO values in winter (e.g., Kalispell, Montana, and Denver, Colorado) can differ greatly in their vulnerability to future exceedances depending on their respective variability in CO concentrations. Of the two Anchorage sites, site 2 (Turnagain) had a lower average CO concentration in winter than did site 1 (1.2 and 2.0 ppm, respectively), but its much greater variability makes site 2 more vulnerable; in fact, the two most recent CO exceedances in Anchorage occurred at this residential site (see Table 1-1). Characterizing the non-Gaussian distribution of 8-hour average CO concentrations in a location might make it possible to predict the probability of future exceedances based on projected emissions inventories. (Gaussian models are discussed in detail in Chapter 3.) In addition, to adequately test the hypothesis that high variability in CO concentrations helps explain the difficulty that some areas have had in meeting the standard, the variability in CO concentrations in areas that met the standard relatively easily should also be examined.

Assessment of Vulnerability

The variability in CO concentrations leads to difficulties in predicting high CO episodes especially in geographical areas with unusually challeng-

---

[11]The ranges of the daily averages for these three meteorological variables for the November through February 1996-1997 winter season were as follows: lower inversion strength, -4.3 to 18.6°C/100 m; windspeed, 0.7-3.9 MPH; and temperature, -43 to +33°F. Data were provided by Paul Rossow of the Fairbanks North Star Borough.

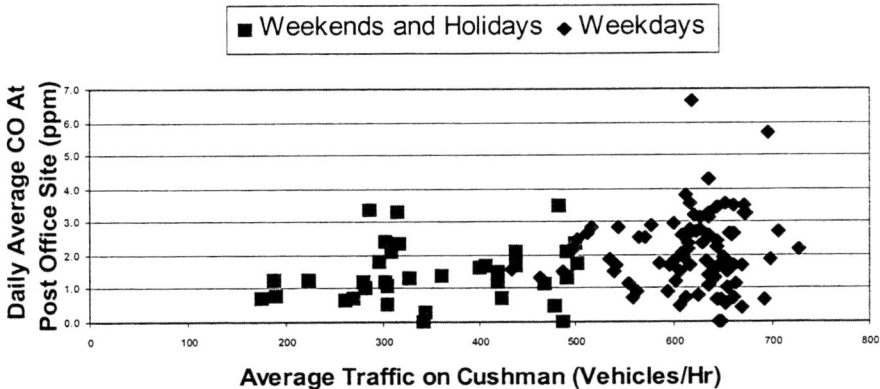

FIGURE 2-11 Scatter plot of Fairbanks traffic and CO during the winter of 1999-2000.

ing meteorological and topographical conditions. Variability in meteorological conditions, such as strong temperature inversions or winds blowing from the direction of nearby communities with high levels of CO, contributes to these difficulties. The status of problem areas could fluctuate between attainment and nonattainment until further emissions reductions provide an adequate safety margin. Areas that have achieved attainment recently and do not yet have an adequate safety margin remain vulnerable to high CO episodes. Nonattainment might occur sporadically under unfavorable meteorological conditions, even when emissions rates remain at the levels projected in the SIP.

Vulnerability can be expressed in terms of the probability of nonattainment in a future year or in terms of the reciprocal of this probability, which can be interpreted as the number of years that are likely to pass until the next nonattainment year takes place. The latter is analogous to a design condition in civil engineering, such as when a bridge is designed to withstand a once-every-hundred-year's flood. It is also analogous to the concept in public health of the number needed to treat (NNT), defined as the reciprocal of the probability for a categorical change in outcome (e.g., from death to survival) for a randomly selected future patient.

Given its stochastic nature, vulnerability is determined by both the central tendency (such as the median or mean) and the spread (such as the

> Recommendations: Vulnerability to Future Violations
>
> Air quality managers typically recognize whether their region is especially vulnerable to future CO violations as a result of increases in vehicle activity, the spatial and temporal variability of meteorology, and problematic topography. However, in some cases, air quality planning does not encompass the worst-case combinations of emissions and meteorology. Achieving sufficient emissions reductions to account for these conditions is prudent, particularly in areas with high population growth and/or high meteorological variability, to further reduce the risk of violations. In addition, given that the form of the CO standard defines a violation as the second and all subsequent exceedances in a calendar year, regions are susceptible to violating the standard due to extreme meteorological conditions, contributing to the difficulties that meteorological and topographical problem areas have in reaching and maintaining attainment. It is also important to investigate how large-scale and local meteorological and climatological phenomena can affect the susceptibility of a location to CO buildup in ambient air.

interquartile range or standard deviation) of the air quality indicator (e.g., the annual second maximum nonoverlapping 8-hour average CO concentration). An area with a large spread might have a substantial probability for nonattainment in future years even after achieving attainment for several years. To reduce vulnerability, emissions reductions must extend beyond the attainment threshold to provide an adequate safety margin.

ILLUSTRATIVE EXAMPLES

This last section provides five illustrative examples of locations that have had problems meeting the NAAQS for CO. The roles of topography and meteorology in concentrating CO and how those factors combine with patterns of emissions to produce episodes of high CO concentrations are briefly described. The committee realizes that the definition of a meteorological and topographical problem area might be somewhat arbitrary, because in all areas meteorology and topography are important factors in producing, concentrating, dispersing, or eliminating all air pollutants. In these examples, the committee primarily focuses on locations in the western United States that experience winter temperature inversions and have topographical features contributing to the accumulation of CO.

Calexico, California

Calexico is located 125 miles (mi) east of San Diego in California's Imperial Valley, on the border with Mexico. This small city's population of 27,109 (in 2000) is similar to Fairbanks's (30,224) (U.S. Census Bureau 2000b).[12] However, it is across the border from Mexicali, a much larger Mexican city with a population of about 750,000. Motor vehicles on the Mexicali side tend to be older and tend to have less sophisticated emissions control equipment that sometimes is not functioning properly or has been removed. In addition, Mexicali has no vehicle emissions inspection and maintenance program. The number of exceedance days recorded since 1995 at a monitoring site in Calexico (59) is surpassed only by the number recorded in Birmingham, Alabama. Although Calexico is a major border crossing point (an estimated 2,168,000 vehicles crossed the border from Mexico in 1999 [Calexico 1999]), CO measurements in Mexicali indicate that the problem is not due to long lines of idling vehicles at the border.

The committee initially thought that CO episodes in Calexico did not fit the profile of locations whose problems were created by meteorology and topography. However, monitoring of CO concentrations in Calexico indicated that the movement of a large, CO-rich air mass northward across the border is responsible for most of the CO pollution in Calexico. This is especially true at night in winter when windspeeds are low and ground-level temperature inversions are strong. The situation is exacerbated by Calexico's topographical location—in a valley down-slope from Mexicali. (The Salton Sea, 30 mi to the north, is 235 feet below sea level.) Although Calexico does not have confining topography that traps or accumulates CO, its topography puts the city in the pathway of CO drifting across the border from Mexicali, and its meteorology prevents CO from dispersing vertically.

Further analysis is needed to assess the relative roles of cross-border transport of vehicles operating in Mexicali compared with vehicles idling at the border in producing CO. To address the problem of cross-border air pollu-

---

[12]The population of the Fairbanks North Star Borough, with a much larger area of about 7,000 square miles, was 82,840 in 2000 (U.S. Census Bureau 2002).

tion, the United States and Mexico have agreed to a new Border Air Quality Strategy, announced by EPA on November 26, 2002 (EPA 2002d).

Lynwood, California

Lynwood is a community of about 64,000 located approximately 12 mi east of the Pacific Ocean. It is south of downtown Los Angeles in the Los Angeles Basin, which has a total population of over 18 million. Lynwood is a densely populated area with numerous freeways and highways, and many high-emitting vehicles. There were 58 exceedance days during the 7-year period from 1995 to 2001.

Numerous studies of the CO problems in Lynwood, including those conducted by the California Air Resources Board (Nininger 1991; Bowen et al. 1996), have been undertaken to assess why CO concentrations are higher in Lynwood than other parts of the Los Angeles area. These studies are aimed to determine the relative contributions of local versus area sources of CO and the roles of meteorology and topography on CO concentrations. Motor-vehicle emissions are clearly important. About half of Lynwood's CO emissions come from just 10% of the light-duty vehicle (LDV) fleet (Lawson et al. 1990). Singer and Harley (1996, 2000) also noted CO emissions rates of vehicles registered in the Lynwood area were double those registered in higher income areas because of the prevalence of older vehicles. Nininger (1991) concluded that the entire Lynwood area is a CO hot spot, due not only to high vehicle emissions but also to lower windspeeds and mixing volumes that occur in surrounding areas. Bowen et al. (1996) further qualified the role of meteorology and topography in contributing to the high concentrations of CO in Lynwood. A strong, surface-based inversion occurs in the Los Angeles area soon after sunset, and the strength of the inversion appears to be greater near Lynwood. In addition, the gradient of the terrain is smaller near Lynwood than at most other locations in the area, resulting in weaker nocturnal drainage winds. The study concluded that significant CO emissions originating in the Lynwood area are added to an urban air mass with high CO concentrations that is transported into the Lynwood area. These emissions sources combine with stable nighttime meteorological conditions to create high CO concentrations in Lynwood.

Fairbanks, Alaska

Fairbanks, Alaska, is a small city in which topography, meteorology, and emissions conspire to produce air pollution in winter (NRC 2002). The meteorological and topographical characteristics of Fairbanks's air pollution problems are discussed in Bowling (1984, 1986). The city has a population of about 30,000 and is located in central Alaska in the Fairbanks North Star Borough, a sparsely inhabited area of over 7,000 square mi with a total population less than 85,000. The city is a center for government, education, and distribution for the northern part of Alaska.

Fairbanks is sheltered by hills to the west, north, and east and is situated on low ground near the confluence of the Chena and Tanana Rivers. The terrain is open to the south, with the Alaska Range roughly 45 mi away.

The meteorology is extreme continental arctic. Low winter temperatures are combined with unusually strong ground-based inversions and low windspeeds. The warmest point in a vertical temperature sounding is commonly more than a kilometer above the surface, and near-surface inversion strengths often exceed 10°C/100 m. These factors greatly limit the amount of air available to dilute and disperse CO and other pollutants.

Temperatures during the winter months are normally below 20°F—below the limits of federal guidelines for cold-start emissions—so engine starting emissions can be substantially higher than normal. As temperatures drop below 0°F, automobiles become increasingly difficult to start. At temperatures below -20°F (not uncommon in Fairbanks during winter), most people use preheating "plug-ins," because it is nearly impossible to start a vehicle that has not been preheated. Engine preheating reduces cold-start CO emissions, so high ambient CO levels are rare at those low temperatures. Encouraging the use of preheating plug-ins at temperatures between 20 and 0°F, when unheated vehicles can be started but emit large amounts of CO, is the centerpiece of the borough's strategies to reduce CO emissions.

Las Vegas, Nevada

Las Vegas is a rapidly growing city in southern Nevada that had a population of nearly 418,000 in 1999, up from about 260,000 in 1990 (U.S. Census Bureau 2000c). Las Vegas is located in a valley surrounded by mountains: the Spring Mountains to the west, the Pintwater, Desert, Sheep, and Las Vegas Mountains to the north; Frenchman Mountain to the east; and the McCullough and Big Spring Ranges to the south. Automobile and truck traffic go to and through the city 24 hours a day via three major highways. In late fall and throughout winter, cool air drainage winds from the adjacent desert hills flow into the city and pool there, resulting in a local accumulation of CO. This pooling effect has resulted in a total of seven exceedance days recorded at two monitoring sites since 1995. Population growth is expected to continue, so future violations are a serious concern.

Las Vegas undertook a significant CO saturation study to help assess the monitoring network and movement of CO (Ransel 2002). The study extensively augmented the 14 permanent monitoring sites with 63 temporary fixed sites and mobile sampling at over 2,500 locations. The study concluded that the current monitoring network captures the peaks and extent of high CO. It also noted that a tongue of high CO appears to be caused by nocturnal drainage flow that follows the Las Vegas Wash. The study noted that, away from the urban core and effects from transport in the drainage flows, CO levels are relatively low (Ransel 2002).

Denver, Colorado

The city of Denver has a population of 501,700 (recorded in 2000) and is located in the South Platte River Valley, approximately 1 mi above sea level. To the west of Denver is the Front Range of the Central Rocky Mountains, with peaks above 14,000 feet. The Cheyenne Ridge (about 70 mi to the north) and the Palmer Divide (about 25 mi to the south) run east to west and are 1,000-2,000 feet above the plain; they combine with the Front Range to form a three-sided basin in which the city sits. Denver is a major national rail center, and two major interstate highways cross the city. This growing community had hundreds of CO exceedances in the 1970s and 1980s. Since 1995, there have been only two. The decline is a result of local controls (including wood-burning bans), technological improvements in motor vehicles and wood burning stoves, and favorable winter weather patterns that have permitted better ventilation of pollutants. However, the city remains vulnerable to future CO exceedances because of its steady population growth.

The meteorological factors that contribute to elevated CO concentrations in Denver include: persistent light winds at the surface, a ground-level inversion, a lee trough along the foothills, snow cover, and warm air advection aloft (Neff and King 1991; King 1991; Neff 2001; Reddy 2001). Reddy (2001) associated elevated CO concentrations with winds at less than 1 MPH and an effective mixed layer less than 25 to 50 m that lasts for at least 3 hours. Neff (2001) also noted microclimatological factors involved in producing exceedances at the most problematic CO monitor. He noted that extensive shadowing by downtown buildings in the afternoon may exacerbate the trapping of pollutants in the area surrounding the monitor by prolonging the cooling of the surface during the winter (thus intensifying the ground-level inversion) and by blocking winds that could disperse pollutants.

3

Management of Carbon Monoxide Air Quality

The Clean Air Act's mandate to "protect and enhance the quality of the Nations air resources so as to promote the public health and welfare" and subsequent scientific findings by the U.S. Environmental Protection Agency (EPA) served as the basis for the National Ambient Air Quality Standards (NAAQS) for carbon monoxide (CO). Chapter 1 discussed the CO standards, trends in ambient CO, and the studies that were influential in developing the health-based standards. Achieving and maintaining the NAAQS requires monitoring ambient CO, developing emissions inventories, implementing emissions regulations and related controls, and tracking progress. This chapter discusses the primary air quality management elements needed to achieve those objectives—the basic emissions control strategies used to reduce emissions and the monitoring and modeling tools used to characterize and assess the magnitude of the problem.

EMISSIONS CONTROL PROGRAMS

CO emissions control strategies have focused on controlling light-duty vehicle (LDV) emissions. The decline in concentrations noted in the earliest stages of CO management in the 1970s corresponded to the implementation of a major enhancement in control of motor-vehicle emissions. There are four approaches for reducing vehicle emissions: (1) new-vehicle certification programs, (2) fleet-turnover incentives, (3) in-use vehicle control strategies, and (4) transportation control measures (TCMs) (Guensler 1998, 2000). This section discusses vehicle emissions control strategies in more detail. As LDV emissions decrease, nonroad, area, and smaller stationary sources may become critical for controlling CO in some locations. This section concludes with a brief discussion on the regulation of these other sources of CO pollution.

Federal New-Vehicle Emissions Standards[1]

Lowering emissions certification standards on new vehicles has been the largest source of reductions in CO emissions from LDVs. For example, the Alaska Department of Environmental Conservation in its most recent SIP for Fairbanks attributed over 70% of total emissions reductions over the 1995-2001 time period to more stringent federal new-vehicle emissions standards (ADEC 2001). Table 3-1 shows emissions standards for passenger cars and light-duty trucks.[2] CO emissions standards have dropped by over an order of magnitude since their inception—emissions from new passenger cars have fallen from 84 grams per mile (g/mi) before emissions controls were instituted to below the current 3.4 g/mi, which began in 1981. New vehicle technologies offering much better environmental performance

---

[1]Vehicles are certified using the federal test procedure (FTP) and the supplemental federal test procedure (SFTP), which specify the preconditioning a vehicle is to undergo before testing, the laboratory conditions the test is to occur in, and a specified driving cycle to be used. Testing is done at temperatures between 68°F and 86°F. Manufacturers are allowed to certify compliance to the 50,000- or 100,000-mile (mi) standards (11 years or 120,000 mi for heavier trucks weighing more than 5,750 lb) using low-mileage cars and an agreed-upon deterioration assumption. However, vehicles may be recalled if emissions control systems are found to be faulty. Tier 2 emissions standards, which will begin with model year 2004, are 120,000-mi standards.

[2]Light-duty trucks have been categorized for emissions certification purposes as light light-duty trucks having a gross vehicle weight rating (GVWR) <3,750 lb (LDT1) or from 3,750 to 5,750 lb (LDT2), and heavy light-duty trucks having a GVWR from 5,751 to 8,500 lb (LDT3 and LDT4). Trucks with a GVWR greater than 8,500 lb are categorized as heavy-duty vehicles.

TABLE 3-1 Federal Passenger-Car and Light-Truck Exhaust Emissions Standards (g/mi)[a]

Model Year	Passenger Cars			Light Trucks[b]		
	HC	CO	NO_x	HC[b]	CO	NOx
Precontrol[c]	10.6	84.0	4.1			
1968-1971	4.1	34.0	—	8.0	102.0	3.6
1972-1974[c]	3.0	28.0	3.1	8.0	102.0	3.6
1975-1976	1.5	15.0	3.1	2.0	20.0	3.1
1977-1978	1.5	15.0	2.0	2.0	20.0	3.1
1979	1.5	15.0	3.1	1.7	18.0	2.3
1980	0.41	7.0	2.0	1.7	18.0	2.3
Tier 0						
1981-1983	0.41	3.4	1.0	1.7	18.0	2.3
1984-1986	0.41	3.4	1.0	0.8	10.0	2.3
1987-1993	0.41	3.4	1.0	0.8	10.0	2.3
1988-1993	0.41	3.4	1.0	0.8	10.0	1.2
Tier 1 (1994-)						
1994 (100,000-mi standards in parentheses)	0.25 (0.31)	3.4 (4.2)	0.4 (0.6)	0.25	3.4	1.2
1995 (100,000-mi standards in parentheses)	0.25 (0.31)	3.4 (4.2)	0.4 (0.6)	0.25	3.4	0.4
NLEV[c] (100,000-mi standards)						
1999	0.09	4.2	0.3	0.09	4.2	0.3

[a]All standards are for 50,000 mi unless otherwise noted.
[b]Standards before 1988 are for all light-duty trucks. Beginning in 1988, light-duty trucks were separated into two weight classes (1988-1993) and then four weight classes (1994-present). The standards after 1988 are for LDT1, which have a 3,750 lb or less gross vehicle weight (GVW).
[c]The National Low Emissions Vehicle (NLEV) Program introduces California low-emissions cars and light-duty trucks into the Northeast in 1999 and the rest of the country in 2001.
Sources: Davis 1997; Chrysler Corporation 1998.

made these achievements possible. This is in contrast to in-use emissions controls, such as vehicle emissions inspection and maintenance (I/M) programs and oxygenated fuels programs, which do not force the adoption of improved vehicle emissions control technologies and hence reduce vehicle emissions by a much smaller fraction (NRC 1999, 2001).

Recent New-Vehicle Emissions Standards

Federal passenger car CO standards have remained at 3.4 g/mi (50,000-mi standard). However there are myriad regulations that have resulted in reductions in vehicle CO emissions. For example, though Tier 1 standards did not affect passenger-car CO emissions, they reduced CO standards for light-duty trucks. Tailpipe emissions of CO and hydrocarbon (HC) respond similarly to changes in air-fuel ratios, and CO is reduced by many of the same vehicle emissions control technologies as HC. Thus, the more stringent HC standards imposed since 1981 have resulted in concomitant reductions in CO.

In addition to these reduced HC standards that result in reduced CO emissions, there are a number of other changes that directly affect CO emissions. With the introduction of Tier 1 standards in 1994, the durability requirements increased from 50,000 mi to 100,000 mi. The supplemental federal test procedure (SFTP), which is discussed in a subsequent section, controls CO during non-FTP conditions of high acceleration and high speed. Cold-start standards, also discussed in a subsequent section, will limit CO emissions during cold-temperature starts.

The recently finalized Tier 2 regulations will also impact CO emissions. Control of tropospheric (ground-level) ozone (O_3), which is caused principally by the interaction of nitrogen oxides (NO_x), certain reactive volatile organic compounds (VOCs), and sunlight on hot summer days, has been a continuing need. On February 10, 2000, EPA promulgated a new series of vehicle emissions regulations, known as Tier 2, intended to partially address this problem by regulating passenger-car and light-duty truck NO_x emissions. Tier 2 requires each manufacturer to meet a sales-weighted "corporate average NO_x standard" of 0.07 g/mi. Lowering fuel sulfur content, which is discussed in the section on in-use emissions controls, is also an integral part of the Tier 2 strategy.

Table 3-2 lists new emissions limits for NO_x, non-methane organic gases (NMOG), CO, formaldehyde (HCHO), and particulate matter (PM) by "bin." Manufacturers certify their vehicles in these bins, ensuring that these vehicles comply with all emissions levels associated with the bins.

TABLE 3-2 Tier 2 and Interim Non-Tier 2 Full-Useful-Life Exhaust Mass Emissions Standards (g/mi)

Bin Number	NO_x	NMOG	CO	HCHO	PM
11[a,c]	0.9	0.280	7.3	0.032	0.12
10[a,b,d]	0.6	0.156/0.230	4.2/6.4	0.018/0.027	0.08
9[a,b,e]	0.3	0.090/0.180	4.2	0.018	0.06
8[b,f]	0.20	0.125/0.156	4.2	0.018	0.02
7	0.15	0.090	4.2	0.018	0.02
6	0.10	0.090	4.2	0.018	0.01
5	0.07	0.090	4.2	0.018	0.01
4	0.04	0.070	2.1	0.011	0.01
3	0.03	0.055	2.1	0.011	0.01
2	0.02	0.010	2.1	0.004	0.01
1	0.00	0.000	0.0	0.000	0.00

[a]This bin and its corresponding intermediate life bin are deleted at end of 2006 model year (end of 2008 model year for HLDTs and MDPVs).
[b]Higher NMOG, CO, and HCHO values apply for HLDTs and MDPVs only.
[c]This bin is only for MDPVs.
[d]Optional NMOG standard of 0.280 g/mi applies for qualifying LDT4s and qualifying MDPVs only.
[e]Optional NMOG standard of 0.130 g/mi applies for qualifying LDT2s only.
[f]Higher NMOG standard deleted at end of 2008 model year.
Source: 65 Fed. Reg. 28 (2000), p. 6855.

For example, a manufacturer might certify their sport utility vehicle (SUV) in bin 7, their passenger car in bin 5, and an economy car in bin 3. All vehicles must meet the full useful life (which has been raised from 100,000 to 120,000 mi) certification limits for their respective bin. NO_x emissions standards for the three bins would then be sales-weighted and compared with the average NO_x standard of 0.07 g/mi.

Tier 2 regulations also allow manufacturers to trade and bank credits. In years that a manufacturer's corporate average falls below 0.07 g/mi it can generate credits which it can bank and use in years when its corporate average exceeds 0.07 g/mi or it can sell these credits to manufacturers whose corporate average is above 0.07 g/mi.

The technologies relevant to the Tier 2 standards will also have benefits for CO reduction. Since the mid-1980s, modern computer-controlled engines have used electronic fuel injectors rather than carburetors to deliver fuel to cylinders in LDVs and most light-duty trucks. The engine computer system reads the signal from an O_2 sensor in the exhaust system and continuously adjusts the air-fuel ratio. The continuous feedback adjustment of the air-fuel ratio is known as closed-loop control. The feedback provides enough air to burn the fuel while maintaining the optimal catalytic-converter efficiency (referred to as the stoichiometric ratio) for control of CO, HC, and NO_x. Figure 3-1 shows the air-fuel ratio effects on catalyst conversion efficiency.

During hard acceleration and high-speed operations, however, engine computers often use fuel-enrichment strategies to enhance engine performance for short time periods and to protect sensitive engine components from high-temperature damage. Likewise, fuel-enrichment strategies are often used during cold starts. Cold temperature CO standards and the SFTP, which are discussed in the following sections, are intended to further control CO for these conditions. Thus, in modern engines, CO as well as HC emissions are most prominent during enrichment associated with heavy loads, hard accelerations, and cold starts. Enrichment factors are much larger for CO compared with HC (see Figure 3-2) (M. Barth, University of California, Riverside, personal communication, October 30, 2002; Scora et al. 2000). Conditions that produce a 10- to 100-time increase in CO emissions produce a 1- to 10-time increase in HC emissions.

The primary methods for meeting the Tier 2 standards—ensuring stoichiometric engine operation over a broader range of operation and promoting faster catalyst warm-up—will have benefits for CO reductions. As shown in Figure 3-3, the prototype Tier 2 vehicle maintains a stoichiometric air-fuel ratio more effectively than a 1996 vehicle certified to California's low emissions vehicle (LEV) standard.[3] Although the current

---

[3]The CAAA90 authorized California, which has the nation's worst air pollution problems, to impose stricter vehicle emissions standards than those for the rest of the nation. California's low emissions vehicles (LEV) regulations require manufacturers to meet fleet-weighted average emissions lower than those mandated by the federal Tier I regulations beginning with the 1994 model year. The California LEV program includes five progressively more stringent categories: transitional low emissions vehicles (TLEVs), LEV, ultra-low emissions vehicles (ULEVs), super ultra-low emissions vehicles (SULEVs), and zero emissions vehicles (ZEVs).

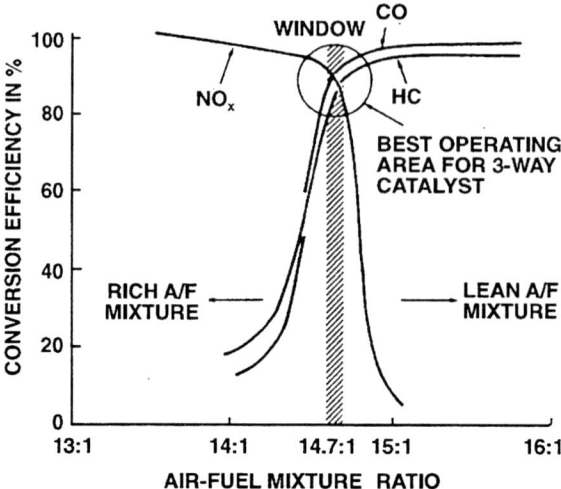

FIGURE 3-1 Catalyst conversion efficiency as a function of air-fuel ratio. Source: Adapted from Canale et al. 1978. Reprinted with permission; copyright 1978, Society of Automotive Engineers.

3.4 g/mi federal new-vehicle standard for passenger cars dates to 1981, the use of advanced-technology three-way catalytic converters and continued improvements in stoichiometric ratio controls have had and will continue to have a collateral CO benefit.

However, it will take years for Tier 2 regulations to be implemented (2007 for LDVs and 2009 for heavier light-duty trucks), and even longer for fleet turnover to occur and for the full benefits of the new technologies to be realized. An increase in vehicle durability has accompanied technological improvements. According to Davis (2001) the national average age of in-use passenger cars has increased from a mean of 5.6 years in 1970 to 8.9 years in 1999. The median lifetime of a 1990 model year passenger car is 4.6 years longer (16.1 years) than that of a 1970 model year car. This increase in vehicle durability will slow the penetration of vehicles with newer emissions control technologies into the fleet. In the meantime, the ongoing improvements resulting from HC standards under Tier 1 and NLEV, the cold-start CO standards, the increased durability required under Tier 1, and the introductions of the SFTP will continue to encourage the downward trend in CO emissions from light-duty vehicles in advance of Tier 2.

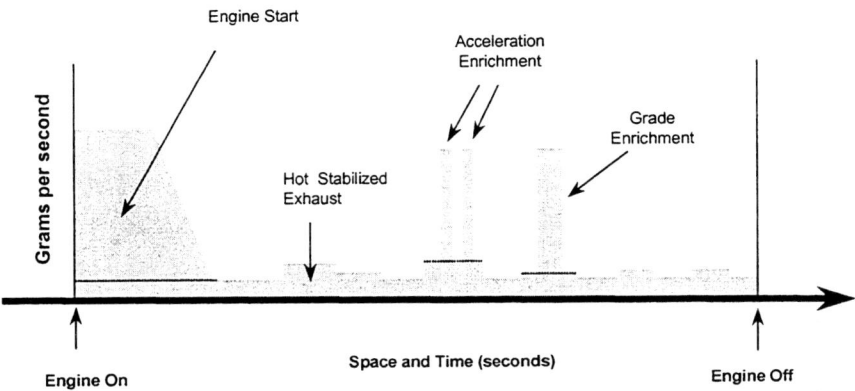

FIGURE 3-2 Hypothetical carbon monoxide emissions rates for typical vehicle operation.

In addition, some emissions control strategies for controlling cold-start emissions are particular to HCs and do not improve CO emissions. Some of the lowest-emitting vehicles in California (called super ultra-low emissions vehicles [SULEV]) use a carbon canister to store uncombusted HC emissions during cold starts. The HC emissions are then recirculated through the catalyst after light-off. Such a control strategy does not reduce CO emissions.

In summary, the impact of Tier 2 requirements is complex. The CO limits for the higher-emissions vehicles, bins 5-8 (bins 9-11 disappear after 2009), is 4.2 g/mi based on a full useful life of 120,000 mi. On the surface, this limit is essentially the same as the Tier 1 and NLEV limits. However, Tier 2 standards are 120,000-mi standards, which should improve vehicle in-use performance. These bins will apply to passenger cars as well as all categories of light-duty trucks (LDT1-LDT4). This will require that LDT2-LDT4, which under Tier 1 standards have 100,000-mi CO standards from 5.5 g/mi to 7.3 g/mi, meet the current CO standards for passenger cars (4.2 g/mi). At a national level, the result of these bins will be a reduction in CO. If bins 6-8 are used for any vehicle, then bins 1-4 must be used to average NO_x below bin 5. Using the example of a manufacturer certifying their SUV in bin 7 (CO standard at 4.2 g/mi), their passenger car in bin 5 (CO standard at 4.2 g/mi), and their economy car in bin 3 (CO standard at

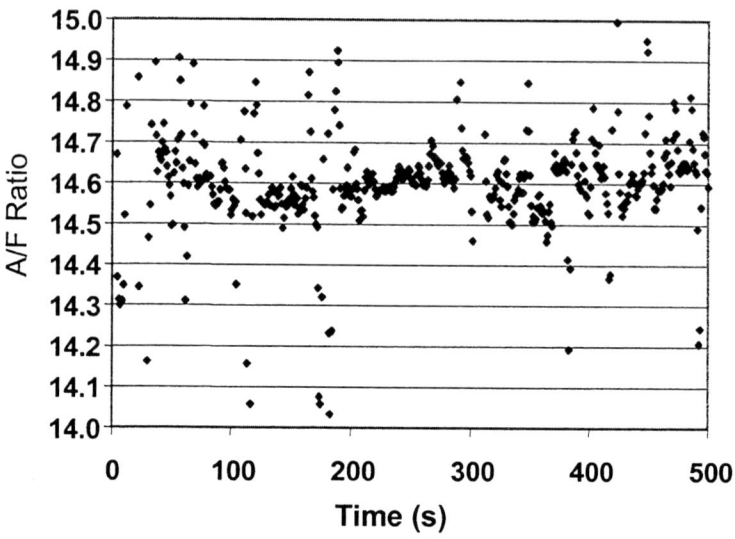

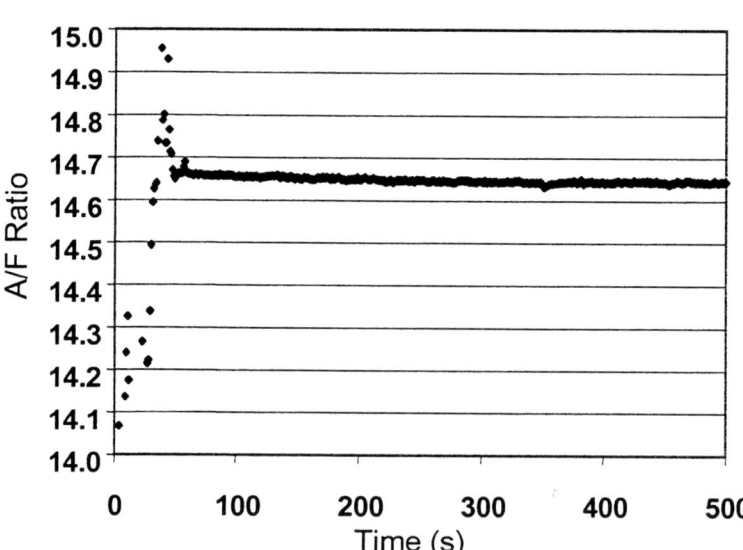

FIGURE 3-3 Example of the improved control of air-fuel ratio resulting from new Tier 2 vehicle technologies. Source: Dana 2002. Reprinted with permission from the author.

2.1 g/mi), the resulting fleet sales-weighted CO will be lower than 4.2 g/mi. However, this is a national fleet average. Local vehicle fleets may differ from the national average.

Cold-Temperature CO Standards

As described in Chapter 1, CO is predominantly a winter problem that occurs in regions known for extreme winter conditions (e.g., Fairbanks, Alaska). During cold starts the engine computer signals the fuel injectors to add excess fuel to the intake air to ensure that enough fuel evaporates to yield a flammable mixture in the engine cylinders. A typical engine-computer strategy injects several times the stoichiometric amount of fuel during the first few engine revolutions, using a fixed fueling schedule to reach idling conditions. Excess fuel continues to be injected until the engine and O_2 sensor are warmed up and the exhaust-catalyst inlet temperature reaches about 250-300°C, sufficient for the catalyst to oxidize CO to CO_2. This open-loop operation, before catalyst light-off (the time it takes the catalyst to reach peak efficiency after start), can continue for several minutes at low ambient temperatures. Cold-start enrichment is responsible for a significant fraction of CO, air toxics, and unburned HCs from properly operating vehicles. Once the engine and emissions control systems are warmed up, combustion becomes stoichiometric, and CO is converted to CO_2 in the catalyst, keeping CO emissions very low under typical operating conditions. Warm up times under mild ambient conditions, at around 70-80°F, can be around 1 min for modern catalysts and even as short as a few seconds for modern close-coupled catalysts (catalysts close to the engine). When ambient temperatures are -20°F or lower, however, catalyst and engine warm-up times can exceed 5 min (Sierra Research 1999). In the case of Fairbanks, Alaska, this means that idling and cold-start emissions from LDVs are particularly high and make up a significant proportion of overall CO emissions. ADEC (2001) and NRC (2002) provide more discussion of the role that cold-start emissions play in Fairbanks.

Since 1994 new cars and the lightest category of light-duty trucks (LDT1) have been required to meet a CO limit of 10 g/mi on the federal test procedure (FTP) cycle conducted at 20°F. For heavier light-duty trucks (trucks between 3,751 and 8,500 lb gross vehicle weight [GVW]), the standard is 12.5 g/mi. The cold-temperature CO emissions standard has been unchanged since it was promulgated in 1994, though certification data from

EPA's certification database show that there have been continued improvement in cold-start emissions (Figure 3-4). Reducing the 10 g/mi limit for the 20°F cold-start test or reducing the test temperature might provide additional CO emissions reductions for cold northern regions, such as Fairbanks. Indeed, CAAA90 mandated that if six or more areas were designated as nonattainment as of July 1, 1997, EPA must require cars to meet a Phase II cold-start emissions limit of 3.4 g/mi. In their presentations to the committee, representatives of the State of Alaska and the Fairbanks North Star Borough discussed how the adoption of the Phase II cold-start standard would aid in Fairbanks's effort toward long-term attainment of the CO NAAQS (Hargesheimer 2001; King 2001; Verrelli 2001). However, EPA has yet to formally determine the number of CO nonattainment areas that existed as of the deadline.

Supplemental Federal Test Procedure

An additional source of CO reductions is the SFTP. The technical community has long known about the absence of high speeds and accelerations from the FTP. The SFTP introduces speeds as high as 80 MPH and maximum accelerations of 8.4 MPH/s into the certification test (61 Fed. Reg. 54852 [1996]). The FTP tests at a maximum speed of 57 MPH and a maximum acceleration of 3.3 MPH/s. Certification to this new cycle will be phased in during the 2000 and 2004 model years. This test procedure should ensure that vehicle emissions control systems will provide improved emissions control over a wider range of vehicle speeds and loads. Much of the improved emissions control will come from reduced use of fuel-rich mixtures at higher loads. EPA estimates a CO emissions reduction of 11% from the LDV fleet in 2020 as a result of the SFTP (EPA 1996). However, it should be noted that for some locations with severe winter driving conditions, such as Fairbanks, the high speed/high acceleration driving conditions within the SFTP are not considered representative. Thus, the benefits from certifying vehicles to the SFTP may be smaller there.

Mobile-Source Compliance Programs

Mobile-source compliance programs are intended to ensure that vehicles meet emissions standards throughout their useful life. There are three

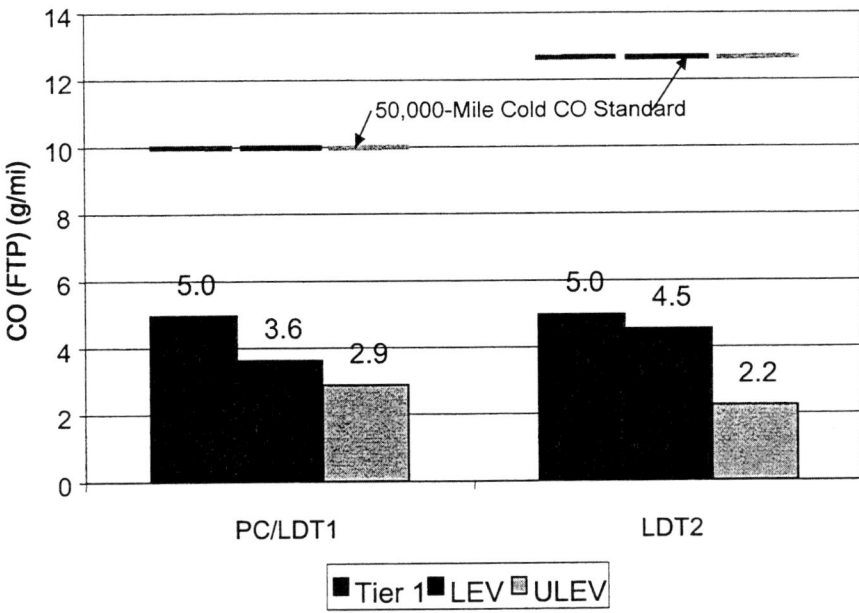

FIGURE 3-4 Average cold CO standards versus certification data (20°F) for model years 2000-2002 at 50,000 mi. There are 279 vehicles represented in PC/LDT1 data (Tier 1 = 91; LEV = 166; ULEV = 22) and 109 vehicles represented in LDT2 data (Tier 1 = 23; LEV = 71; ULEV = 16). Source: Dana 2002. Reprinted with permission from the author.

major components of the program: preproduction evaluation, production evaluation, and in-use evaluation. Before a motor vehicle can be sold in the United States, EPA requires that it be emissions certified. This involves testing a preproduction prototype to prove that it meets applicable model year emissions standards and durability-deterioration requirements. The certificate requires the manufacturer build every vehicle (or engine) the same as the prototype in all material respects. Once production begins, the manufacturer and EPA conduct end-of-line production audits of randomly selected vehicles. These audits frequently require emissions tests to determine conformity with applicable standards. If a statistically significant number of vehicles fail, the manufacturer takes steps to repair them. Failures may involve a faulty or out-of-specification part or adjustment or improper assembly. Repair measures can include stopping production, redesigning, repairing vehicles already produced, and recalling those that have been sold. In addition, EPA uses data from state vehicle emissions inspections and maintenance (I/M) programs, technical service bulletins, and voluntary emissions recall reports to target testing of in-use vehicles so that in-use compliance can be verified. Recently, manufacturers have been required to develop a general program for in-use testing of customer-owned vehicles. The committee is unaware of any in-use cold-temperature compliance testing and is concerned about the lack of such programs.

Motor-Vehicle In-Use Controls

While tougher emissions certification standards are the primary means by which CO emissions have been reduced over the last three decades, in-use vehicle controls have provided additional reductions and ensured vehicles are properly maintained. Clean fuels programs and I/M programs are the mainstays of nationwide in-use vehicle controls. In some regions, where winter temperatures are frequently below 20°F and cold-starts contribute substantially to the regional emissions inventory, engine preheating also provides valuable emissions reductions.

Vehicle Emissions Inspection and Maintenance Programs

Vehicle emissions I/M programs are designed to identify vehicles that have higher than allowable emissions and ensure that they are repaired or removed from the fleet. They are the most common measure implemented by state and local governments to reduce CO emissions. I/M programs attempt to control emissions throughout a vehicle's lifetime by ensuring that the vehicle's emissions control system is maintained and repaired when necessary. These programs are implemented in nonattainment areas and in other areas seeking to improve air quality. The inspection traditionally involves regularly scheduled exhaust tests administered at a certified testing facility. The test measures CO, HC, and sometimes NO_x emissions. I/M tests also include a visual inspection of the components controlling evaporative and exhaust emissions and may include a functional gas-cap test and a pressure test of the evaporative emissions control system. New testing technologies, such as those using onboard diagnostic (OBD) systems and remote sensing, are also being used.

The three basic types of I/M programs currently in operation are centralized, decentralized, and hybrid. A centralized program consists of a relatively small number (relative to a decentralized network) of stations that perform only emissions tests. Vehicles that fail the inspection must be repaired elsewhere. This program typically is operated by a government entity or by a contractor with government administration. A decentralized testing program consists of a larger number of low-volume stations that do both emissions testing and vehicle repairs. This type of program links testing to the repair process and is operated by private sector stations. Finally, a hybrid program is one that incorporates elements of both decentralized and centralized programs (NRC 2001).

Traditional Exhaust Emissions Tests

I/M programs began with an idle test for HC and CO. This test uses a tailpipe probe to measure the steady-state concentrations of CO, HC, and carbon dioxide (CO_2) emitted from idling vehicles. A high-idle test in which engine speed is manually increased to approximately 2,500 revolutions per minute (rpm) is sometimes performed in addition to the natural or "low-idle" test; in traditional idle testing, there is no load applied to the engine.[4] NO_x concentrations are not measured during idle tests because NO_x emissions are low during no-load idle conditions.

Because the idle test cannot assess the performance of NO_x control technologies, it has been superseded by the IM240 and ASM tests. The IM240 is a loaded-mode transient dynamometer test that measures the mass of CO, HC, and NO_x emissions collected over a simulated 240-second (s), 2-mi driving cycle. The ASM (acceleration simulation mode) test involves the acceleration of a vehicle on the dynamometer to a steady-state speed while measuring exhaust concentrations of CO, HC, and NO_x.

More than 30 states now operate I/M programs with a wide assortment of test designs. An NRC committee that recently looked at I/M programs concluded that they generally have not achieved the emissions reductions originally projected (NRC 2001). For example, in-program and remote-

---

[4]A loaded-mode test, such as the IM240 or the FTP, involves testing vehicle emissions while the vehicle is on a dynamometer that simulates the load a vehicle is under during on-road operation.

sensing data estimated that the Colorado I/M program achieved reductions of 4-11% compared with a modeled estimated benefit of 17-34% (Stedman et al. 1997, 1998; ENVIRON 1998; Air Improvement Resources 1999). In addition, there is a need to assess the emissions reduction benefits of I/M for vehicles operating in cold temperatures. However, that committee identified a great need for continuing programs that repair or eliminate high-emitting vehicles from the fleet given the major influence of this small fraction of the fleet on total emissions and air quality. The NRC (2001) report estimates that the dirtiest 10% of LDVs produce 50-60% of on-road LDV exhaust emissions. The report also discusses studies that combine data for vehicle ownership, high-emitter frequency, and income levels. These studies suggest a strong link between low household income and the likelihood of owning a high-emitting vehicle. Because of this link, the NRC (2001) report recommended that the cost-effectiveness and equity impacts of such policies be explored to provide financial or other incentives, such as repair assistance programs, for motorists of high-emitting vehicles to seek repairs or vehicle replacement.

Onboard Diagnostic Systems

New technologies providing faster, more convenient, and more accurate emissions testing continue to be developed. Many make use of the onboard diagnostics (OBD) systems in vehicles. The enhanced onboard diagnostics system installed in model year 1996 and newer vehicles, known as OBDII, can help to detect problems that increase CO emissions. OBDII uses sensors to monitor and modify the performance of the engine and emissions control components. The onboard computer monitors signals from the sensors and actuators to identify sensor and control-system failures, illuminating the malfunction indicator light (MIL) on the vehicle dashboard, and storing the fault codes (known as diagnostic trouble codes) for later analysis. In a garage setting, mechanics can download the OBDII fault codes from the onboard computer with a diagnostic analyzer or "scan tool." The codes identify emissions control systems and components that are malfunctioning. It should be noted that OBDII is only a diagnostic system and does not, in the absence of an OBD I/M program, require a vehicle owner to repair an emissions-related problem detected by the system.

If an OBD I/M program is operating properly, OBDII inspections should fail vehicles if the vehicle's emissions control components are or have been malfunctioning or if the sensors monitoring emissions control components are malfunctioning. In contrast, traditional I/M emissions-testing programs inspect actual vehicle emissions for violations of standards set by individual states. It should be noted that vehicles equipped with OBDII have the ability to maintain low emissions (relative to older technology vehicles) even after a system component has failed.

EPA (65 Fed. Reg. 56844 [2001]) recently finalized an OBD rule that requires states to implement OBD testing in I/M programs for 1996 and newer OBD-equipped vehicles. However, the proposed replacement of traditional emissions-testing programs with OBDII programs has been controversial. The major issues, identified in the NRC (2001) study and by EPA's Mobile Source Technical Review Subcommittee (EPA 2002e), include the following: (1) the lack of overlap in some studies of vehicles that fail both the OBD I/M tests and traditional tailpipe emissions tests; and (2) the significant fraction of vehicles that failed OBD I/M tests with actual emissions below the vehicle's certification standards.

In addition, some components of OBDII systems (such as exhaust-gas recirculation and O_2 sensors) are often disabled by the engine computer during conditions under which the manufacturer cannot guarantee the components' performance (J. Cabaniss, Association of International Automobile Manufacturers, personal communication, July 10, 2001). That tends to be the case for vehicles operating at temperatures below 20°F. There is a need to understand the behavior and performance of OBDII at low temperatures, especially if many northern locations begin adopting OBD I/M systems. When a significant number of sensors become inoperative, the OBDII system's ability to alert vehicle owners of potential emissions-system failures is diminished.

Remote Sensing

Remote sensing has also been used as a new I/M testing method. Remote sensing is a technique that measures emissions from individual vehicles as they drive by a roadside sensor. It offers the possibility of testing vehicle exhaust emissions without requiring the vehicle's presence at a testing facility, though the test is only for a relatively short time (approximately ~0.5 seconds) during which the remote-sensing beam passes through the exhaust. Remote sensing is most accurate for measuring CO. It is currently being used in Colorado and Missouri to identify clean vehicles

> **Recommendations: High-Emitting Vehicles**
>
> Air quality management agencies should identify high-emitting vehicles and target them for repair or removal from the fleet. Enhanced onboard diagnostic testing programs, tailpipe testing, motor-vehicle emissions profiling, and/or remote sensing can identify these vehicles. However, programs designed to mandate repair or removal of high-emitting vehicles might raise issues of fairness, because high emitters are often owned by people with limited economic means. Policies that provide incentives for owners of high-emitting vehicles to seek repairs or vehicle replacement, such as repair assistance programs, should be explored. There should also be additional low-temperature testing to evaluate the effectiveness of programs aimed at controlling high-emitting vehicles. This evaluation should include the impacts on CO as well as other emissions.

that may opt to avoid visiting an emissions testing station for scheduled testing. For example, in the St. Louis area, if a vehicle has two or more successive low-emissions readings measured by remote sensing, the vehicle owner can opt to be excused from scheduled emissions testing.

The implementation of remote sensing for identifying high emitters in Arizona, however, was terminated after 5 years by state legislators because of problems including high costs, false failures, and difficulties finding appropriate remote-sensing sites. From mid-May 1998 through early June 1999, over 2 million valid remote-sensing test records were collected, but only 2,987 vehicles were identified as high emitters (Wrona 1999). Owners of vehicles identified as high emitters were sent letters ordering them to submit their vehicles for IM240 testing within 30 days. About half (55%) of vehicle owners responded within that time period; 15-20% of vehicle owners complied later, after their vehicle registration was suspended. Of vehicles that reported for testing, 42% passed the initial IM240 test. A survey indicated that one-third of those vehicles underwent repairs prior to the test (Wrona 1999). Other vehicles may have been repaired, but owners may not have reported that on the survey.

Besides application as a testing device, remote-sensing measurements can be useful for characterizing vehicle emissions, including average emissions by model year and the fraction of high emitters in the vehicle fleet. Remote sensing can also help assess I/M program effectiveness and estimate the extent of certain types of program noncompliance.

Fuels

Vehicle emissions control has also occurred through fuels reformulation. This includes increasing the oxygen content of fuels to promote more complete fuel combustion, reducing sulfur content to improve catalyst efficiency, and switching to fuels that inherently produce less CO during vehicle operations.

Oxygenated Fuels

In 1988, the use of oxygenated fuels (or oxyfuels) was instituted in Colorado to reduce winter CO levels and was subsequently extended to other areas of the United States that were exceeding the NAAQS for CO (typically during winter). EPA mandated that oxyfuels contain an oxygenate (normally either methyl *tertiary*-butyl ether [MTBE] or ethanol) with oxygen content of 2.7% or more by weight.

Adding an oxygenate to the gasoline increases the oxygen-fuel ratio in the combustion process, changing the combustion chemistry and decreasing the emissions of CO formed during incomplete combustion. A 1997 study of the winter oxyfuels program initiated by the White House Office of Science and Technology Policy concluded that at temperatures above 50°F, CO emissions from most vehicles were reduced by about 3-6% per weight percent oxygen (NSTC 1997). CO emissions reductions of 3-7% are predicted by EPA's MOBILE6 model for the 2010-2015 fleet, mainly because of reduced emissions from pre-1994 vehicles, cold starts, and malfunctioning vehicles.[5] Emissions reductions are generally lower in newer-technology vehicles (those with closed-loop fuel control and three-way catalysts) and higher in high-emitting, older-technology vehicles (those with permanent open-loop fuel control and two-way catalysts). O_2 sensors and on-board computers in later models control the air-fuel ratio to prevent fuel-rich operations.

---

[5]Earlier versions of the MOBILE model predicted much larger benefits from oxyfuels. In a review of the winter oxyfuel program, the Office of Science and Technology Policy found that the observable reduction in ambient CO levels that could be attributed to the use of fuel oxygenates was lower than the amount predicted by the MOBILE5a model by a factor of 2 or 3 (NSTC 1997).

There is a lack of information on the effectiveness of oxyfuels at temperatures below 50°F. EPA (Mulawa et al. 1997) tested three vehicles at 20°F, 0°F, and -20°F at its cold-weather facility using unleaded gasoline containing 10% ethanol (3.5% oxygen by weight). Two of the cars showed substantial improvement in CO emissions; the third showed none. The Colorado Department of Public Health and Environment found an average 11% decrease in CO emissions by switching to 10% ethanol blended fuels in 24 vehicles that it tested at 35°F (Ragazzi and Nelson 1999). One of the problems with these oxyfuel studies is that the vehicles used might not be operating the most current control systems and might be even less representative of Tier 2 vehicle technology. Theory suggests that oxygenated fuels would provide emissions benefits under extreme cold-start conditions because cars run under open-loop conditions for longer periods. However, available data are not sufficient to support or refute that argument. Because CO has become less of a problem in many places, the number of new studies looking at the effectiveness of oxyfuels at low temperatures has decreased considerably.

Although oxyfuels provide some air quality benefits, concerns have been raised about the widespread use of MTBE as an oxygenate. The offensive odor and taste of MTBE, and the potential adverse effects of MTBE leaking into drinking water supplies have raised questions about whether the benefits gained from using MTBE (reducing high ambient CO concentrations 1 or 2 days per year) are greater than the possible negative consequences.

Reformulated Gasolines

Reformulated gasoline (RFG) is mandated by EPA for use in areas exceeding the NAAQS for ozone. RFGs must meet a number of requirements both in fuel composition (benzene < 1.0% and oxygen > 2.0% by weight) and in reduction of exhaust emissions of VOCs and air toxics, including benzene. In practice, this translates into an RFG aromatic content of <25%, and fuel sulfur concentrations of about 30 ppm (compared to >350 ppm for non-RFGs). Kirchstetter et al. (1999) concluded that the use of reformulated gasolines in California led to reduced CO emissions from hot-stabilized vehicles, consistent with the body of data from the Auto/Oil Air Quality Improvement Research Program (NRC 1999). The decrease in CO emissions associated with RFGs is in part due to their lower sulfur content compared with regular gasolines (NRC 1999).

Low-Sulfur Fuel

A key finding of the Auto/Oil project was that reducing fuel sulfur decreases exhaust emissions (Benson et al. 1991). Sulfur in gasoline adversely affects the efficiency of vehicle emissions control systems by poisoning the catalyst. This decreases pollutant conversion efficiency and potentially lengthens the time needed after ignition for the catalyst to become effective. The 1991 Auto/Oil study concluded that reducing sulfur concentrations from 450 ppm to 50 ppm would result in a 13% decrease in CO exhaust emissions in 1990 Tier 0 technology vehicles. In addition, low-sulfur fuel is expected to lengthen lubricant and engine life as well as reduce emissions of HC, NO_x, hydrogen sulfide, sulfur dioxide, sulfuric acid aerosols, and other air toxics. Reversing the effects of sulfur on catalytic performance requires fuel-rich conditions and aggressive accelerations that achieve high catalyst temperatures (about 1,200°F). However, sulfur's effects are not easily reversed in the newer-model lower-emissions vehicles (Truex 1999). To guard against the poisoning effects of sulfur, it is best to operate these newer-model vehicles on low-sulfur fuel only.

To address concerns about the increased sensitivity of the newer-technology vehicles to sulfur poisoning, EPA included new fuel standards requiring refiners to meet an average sulfur concentration of 30 ppm beginning January 1, 2006, in its Tier 2 proposals (65 Fed. Reg. 6697 [2000]). However, additional studies on the effect of high-sulfur gasoline on catalyst efficiency and light-off time in cold climates are necessary.

Hybrid Gasoline-Electric Vehicles

Hybrid vehicles combine a conventional internal combustion engine with an electric motor. The internal combustion engine can be run on various alternative fuels; however, this discussion relates to those vehicles powered by a normal gasoline-fueled engine alternating or in concert with an electric motor, with the battery system being charged by the gasoline-powered engine. Compared with a regular gasoline-fueled vehicle, this arrangement allows greater gasoline mileage to be achieved while maintaining low exhaust emissions. Hence, hybrid vehicle emissions are comparable to emissions from gasoline-fueled vehicles with the same emissions rating (see DOE/EPA [2002] for emission ratings of available hybrid vehicles). The committee was unable to find emissions data for hybrid vehicles under cold-weather temperature conditions like those encountered in Fairbanks. However, the extreme cold temperatures may make use of the electric motor/batteries infeasible.

Compressed Natural Gas and Liquified Petroleum Gas

Vehicles fueled by compressed natural gas (CNG) and liquefied petroleum gas (LPG) are inherently cleaner, at least in terms of reactive HC emissions and ozone formation, than gasoline-fueled vehicles. As shown in Table 3-3, CO emissions, estimated using the FTP from five CNG-fueled vehicles (one 1999 passenger car and four 1994-1995 pick-up/light-duty trucks) and three LPG-fueled trucks, were significantly lower than the federal CO emissions standard. However, it is unknown how these vehicle's emissions rates will deteriorate with increasing mileage.

Collateral Emissions Reductions from Emissions Standards and In-Use Controls

As noted in Chapter 1, CO controls can reduce emissions of other pollutants generated during fuel-rich or cold-start engine operation, such as PM, benzene, 1,3-butadiene, polyaromatic hydrocarbons (PAHs), and aldehydes. A recent assessment by EPA found that one of the major sources of air toxics exposure nationally is automobile emissions (EPA 2000c). The agency recently finalized regulations on the control of air-toxics emissions from mobile sources (66 Fed. Reg. 17230 [2001]). The EPA analyses show that programs already in place, such as the reformulated gasoline program, the national low-emissions vehicle program, and the Tier 2 emissions standards and fuel sulfur controls, will yield significant reductions of mobile-source air toxics. Table 3-4 displays emissions reductions estimated to occur due to existing federal programs for four selected mobile-source air toxics.

Quantifying air toxics emissions reductions from fuel reformulation is rather uncertain at present. The primary tool used for that assessment is the COMPLEX model from EPA. It is based on a rather limited range of automobile control technologies and temperatures (NRC 2000) and should be updated to provide better estimates for emissions responses for air toxics, including CO's response across a range of relevant conditions. Model im-

TABLE 3-3 Summary of FTP Emissions Results for the Test Fleet in the AFV Study

Model Year	Vehicle	Fuel	CO (g/mi)
1999	Honda Civic GX	CNG	0.026
1995	GMC Sonoma PU	CNG	0.977
1994	Dodge Caravan Minivan	CNG	0.200
1994	Dodge Ram 350 van	CNG	0.913
1994	Dodge Ram 350 van	CNG	0.217
2000	Ford F-150 XL	LPG	0.145
1999	Ford F-250 XLT	LPG	0.420
1992	Chevrolet S10 PU	LPG	0.492

Abbreviations: AFV, alternative fuel vehicles; CNG, condensed natural gas; FTP, federal test procedure; GMC, General Motors Corporation; LPG, liquified propane gas.
Source: J. Norbeck, University of California, Riverside, unpublished material, 2002.

provements will improve the assessment of the collateral impact of CO emissions reductions on air toxics.

It should be noted that some CO control technologies might lead to increased emissions of other pollutants. For example, oxygenates can increase NO_x emissions, which in turn can increase concentrations of both ozone and particulate nitrate (NRC 1999). In addition, the use of oxygenated fuels for control of CO is expected to decrease emissions levels of PAHs and benzene but increase the levels of certain emitted aldehydes. Thus, the picture is complex and strongly argues for an integrated approach to air quality management that does not isolate pollutants.

Transportation Control Measures

Transportation control measures (TCMs) are actions designed to change travel demand or vehicle operation characteristics to reduce motor-vehicle emissions, energy consumption, and traffic congestion. Transportation agencies are increasingly experimenting with new TCMs, some of which are listed in Table 3-5. TCMs include transportation-supply improvement (TSI) strategies and transportation-demand management (TDM)

TABLE 3-4 Estimated Percent Reduction for Selected Toxics (for Nationwide On-Highway Vehicles Only)

Compound	Cumulative Percent Reduction from 1990		
	1996	2002	2007
Benzene	33%	65%	73%
Formaldehyde	33%	69%	76%
Acetaldehyde	23%	58%	67%
1,3-Butadiene	35%	67%	72%

Source: 65 Fed. Reg. 6697 [2000].

strategies. TSI strategies attempt to reduce emissions by changing the physical infrastructure of the road system to improve traffic flow and to reduce stop and go movements. In contrast, TDM strategies attempt to reduce the frequency and length of automobile trips by changing driver behavior using regulatory mandates, economic incentives, voluntary programs, and education campaigns.

TSI strategies can be grouped into three categories: traffic signalization, traffic operations, and enforcement and management. Traffic signalization strategies include programs to optimize the timing of individual traffic signals and to coordinate traffic signals over a designated area. Traffic operations strategies include converting two-way streets to one-way streets, restricting left turns, "channelizing" roadways and intersections, and selectively widening roadways and intersections to reduce bottlenecks. Enforcement and management strategies include incident management programs to reduce delays from accidents and other roadway incidents, ramp metering to improve the flow of traffic on freeways, and general enforcement of traffic and parking regulations. These techniques have been used for decades and are considered cost-effective strategies for reducing congestion, but their effects on vehicle emissions are difficult to measure and predict (EPA 1998b). The Federal Highway Administration (FHWA) has described methods for estimating the emission reductions of many TCMs for the Washington, D.C., area (FHWA 1995).

Demand-management measures include, but are not limited to, no drive days, employer-based trip-reduction programs, parking management, park and ride programs, work-schedule changes, transit-fare subsidies, and public-awareness programs. These measures fall into four categories: trip-reduction mandates, market incentives, voluntary programs, and education

TABLE 3-5 Transportation Control Measures (TCMs)

IMPROVED PUBLIC TRANSIT
Incentives for single occupancy vehicle commuters to use convenient and reasonably priced mass transit alternatives. The three major ways of increasing ridership on public transit are (1) system/service expansion, (2) system/service operational improvements, and (3) inducements to increase ridership.
TRAFFIC FLOW IMPROVEMENTS
Strategies that enhance the efficiency of a roadway system, without adding capacity, including traffic signalization, traffic operations, and enforcement and management.
HIGH OCCUPANCY VEHICLE (HOV) LANES
Roadways dedicated for HOV use.
INTELLIGENT TRANSPORTATION SYSTEMS
Traffic detection and monitoring, communications, and control systems. Examples include traffic signal control, freeway and transit management, and electronic toll collection systems.
BICYCLE AND PEDESTRIAN PROGRAMS
Includes sidewalks, bicycle lanes, and bicycle racks.
COMMUTE ALTERNATIVE INCENTIVES
Incentives, usually employer based, to encourage commuters to carpool or use transit services.
TELECOMMUTING
Working at home using electronic communication instead of physically traveling to a distant work site.
GUARANTEED RIDE HOME PROGRAMS
Ensures transportation (e.g., taxi or transit passes) for carpooling employees in the case of an unforeseen circumstance.
WORK SCHEDULE CHANGES
Adjusting hours worked to control peak emissions. Examples include staggered hours, flextime, and a compressed workweek.
TRIP REDUCTION ORDINANCES (REGULATORY MANDATES)
Regulations that attempt to adjust personal travel decisions through employer-based incentive/disincentive programs.
CONGESTION PRICING
Financial disincentives to driving on highly used roadways, or priced alternatives to a congested roadway. *(Continued)*

TABLE 3-5 *Continued*

PARKING PRICING

Programs that encourage single-occupant vehicle users to switch to other means of travel by imposing fees for parking or that encourage shifting times for vehicle starts away from peak CO periods.

PARKING MANAGEMENT

Allocation of parking spaces intended to encourage single-occupant vehicle users to use other means of travel.

Source: EPA 2001g.

and exhortation campaigns (Guensler 1998). In general, trip-reduction mandates, such as trip-reduction ordinances that require employers to establish demand-management programs, have not proved effective in the United States (Guensler 1998). However, these mandates may be effective for very large employers, such as universities and major government centers. Market incentives, including transit-fare subsidies, may offer greater potential. Various strategies for increasing the direct cost of driving—market disincentives—have also been proposed but are rarely implemented. The success of voluntary control strategies depends on consumer behavior and the availability of alternatives, so public education and exhortation programs figure prominently in all of these strategies.

In a recent review of the Congestion Mitigation and Air Quality Improvement Program (CMAQ), which funds TCMs in ozone and CO nonattainment areas, the Transportation Research Board (TRB 2002) found broad support for the program among transportation planners, air quality officials, and interest groups. Although the limited evidence available suggested that TCMs were less effective in terms of costs per emissions reduced compared with other emissions reduction strategies, the program offers the opportunity for nonattainment areas to experiment with nontraditional transportation approaches to pollution control. In addition, CMAQ also funds some promising TCMs that receive limited if any support from traditional transportation funding sources.

Most TCMs have been developed for large metropolitan areas and for areas in nonattainment for ozone; TCMs for Los Angeles were described by Bae (1993). Smaller regions and regions facing CO problems must adapt these TCMs or develop ones specific to their needs. In Fairbanks, Alaska, for example, CO emissions in winter are substantially increased by cold starts. The Fairbanks North Star Borough has adopted a "plug-in" program as one of its primary transportation control measures. Vehicles do not start readily at temperatures below 0°F, so residents install electric engine block heaters to keep their engines warm when parked for extended periods of time. The borough's plug-in program involves two components. The first is a public education campaign to encourage residents to plug in their vehicles at temperatures from 0°F to 20°F, when CO emissions are high, even though vehicles can start without being plugged in. The second component is an ordinance requiring large employers to install electric outlets in their parking lots (NRC 2002).

Public Education Programs

Public education programs are designed to increase public awareness and understanding of air quality problems and may lead to changes in behavior that result in emissions reductions. Available evidence suggests that public awareness and understanding levels of air quality problems are low. A study conducted by the U.S. Department of Transportation and EPA as a part of the Transportation and Air Quality Public Information Initiative concluded that citizens do not understand the link between transportation choices and air quality, are largely unaware of the range of alternatives to solo driving available in their communities, and do not place a high priority on air quality and transportation issues (DOT/EPA 2002). For example, although Fairbanks has a fairly active public-information campaign concerning the connection between vehicle plug-ins at temperatures above 0°F and improved air quality, most individuals responding to a survey said they plugged in for ease in starting their vehicles (ADEC 2001).

Public education programs have been implemented in numerous metropolitan areas throughout the United States by local, regional, and state governments as well as nonprofit organizations such as the American Lung Association, with help from the federal government. In May 1999, the FHWA, Federal Transit Administration (FTA), and EPA's Office of Transportation and Air Quality developed the "It all adds up to cleaner air" program (DOT/EPA 2003). This program provided federal support in the form of market research, advertising, a "Comprehensive Resource Toolkit," an orientation workshop, and limited funding for 14 demonstration communities and provided materials related to public education to many others. However, the cost-effectiveness of public education programs has not been documented.

Episodic control programs aim to change travel and other kinds of behavior on days when exceedances of air quality standards are possible. These programs, also called "action day" programs, largely depend on public service announcements and other forms of public education but may also involve incentives to change behavior (e.g., free transit fares). As of 1996, at least 35 regions in the United States had implemented or were developing episodic control programs, in maintenance areas as well as nonattainment areas (EPA 1997b), and by 2002, the number had grown to well over 50 (EPA 2002f). Most of these programs target ozone, a more pervasive problem than CO, but the Air Pollution Control Division (APCD) of the Colorado Department of Public Health and Environment issues advisories for CO and PM in winter months that activate mandatory woodburning restrictions and call for voluntary driving reductions (Regional Air Quality Council 2002). Rigorous evaluation of the effectiveness of these programs is not available, but EPA has said that episodic controls "have the potential of being more effective in reducing short-term air quality violations" than long-term emissions-reduction measures (EPA 1997c). In addition, these programs may be more acceptable to the public than long-term restrictions on driving or gasoline use (EPA 1997b).

In order to implement an episodic control program, a regional air quality agency must be able to predict when conditions will be conducive to an exceedance. In most programs, alerts are issued the day before and depend primarily on weather conditions, including winds, temperature, and cloud cover. An ability to predict exceedances with perfect accuracy is not essential, but calling too many alerts is likely to reduce the effectiveness of the program. As discussed in the committee's interim report (NRC 2002), the Fairbanks North Star Borough has called 16 alerts over the past four winters; four were correct (an exceedance actually occurred), and 12 were not.[6] During the same period, three exceedances occurred that were not forecast.

In 1997, EPA established a policy for incorporating voluntary measures, such as public education programs and episodic control programs, into state implementation plans (SIPs) and giving SIP credits for them. Called the Voluntary Mobile Source Emission Reduction Program, its aim is to make it easier for state and local governments to achieve air quality

---

[6]While it is possible that emissions were reduced in response to broadcast alerts, a preliminary analysis of vehicle traffic does not indicate significantly less driving on alert days.

> **Recommendations: Public Education**
>
> Public-health education to improve public acceptance and compliance should be a component of all local emissions-reduction programs. Communities should use surveys and focus groups to regularly evaluate the effectiveness of public education programs and the impact they have on the success of CO emissions control.

goals by providing greater flexibility in determining the best measures for their communities. The policy allows as much as 3% of the total reductions needed for attainment to be from voluntary mobile-source programs. To claim credit, states must provide a realistic estimate of the emissions impact and commit to monitoring the success of the program and remedying any shortcomings (EPA 1997d).

Land-Use, Urban Population Growth, and Sprawl

The sprawling patterns of land development typical of metropolitan areas in the United States contribute to high levels of automobile travel and thus to air quality problems. The defining characteristics of "sprawl" include low-density development, unlimited outward expansion, and "leap-frog" development (Burchell et al. 2002). Most metropolitan areas in the United States are growing faster in land area than in population. Between 1982 and 1997, urbanized land increased by 47%, while population grew by only 17% (Fulton et al. 2001). This low-density pattern of growth has two important effects on travel: longer trip distances and greater reliance on the car.

Land-use policies are increasingly recognized as a consideration in formulating an overall strategy to combat congestion and are also now recognized by EPA as a tool for improving air quality. "Smart growth" programs designed to counter sprawl are popular throughout the United States. These programs use both regulations (such as zoning) and financial incentives to encourage development within existing urbanized areas, which is conducive to public transit, biking, and walking. Smart growth strategies have the potential to reduce vehicle travel by reducing trip distances and reliance on the car (EPA 2001f). However, the full impact of such programs is uncertain. In recognition of their potential to reduce emissions, EPA now allows state and local communities to account for the air quality benefits of smart growth strategies in SIPs as a part of the Voluntary Mobile Source Emission Reduction Program (EPA 2001g).

The benefits of smart growth strategies may be more likely to accrue at the regional level than they are at the local level. Smart growth policies may lead to higher densities of development in certain areas within the metropolitan region and thus to potentially higher levels of vehicle traffic in those areas. The increase in traffic could, in turn, lead to higher localized concentrations of CO and other motor-vehicle pollutants. It is thus possible that smart growth strategies will prove effective in reducing regional levels of ozone but at the time result in the creation of new areas of high CO and related pollutants. However, given the continuing reduction in CO emissions through improvements in vehicle controls, the possibility that such areas would produce CO exceedances seems remote for most locations.

Control of Stationary and Area Sources

Virtually any process that burns fossil fuels or biomass will produce CO, though in varying quantities. The more efficient the combustion process, the lower CO emissions will be. Thus, because of high combustion temperatures, power plants and most other industrial processes have very low CO emissions relative to the amount of fuel burned.[7] As was discussed in Chapter 1, the situation in Birmingham, Alabama, is an exception. An industrial source (a mineral-wool production facility) has been the cause of numerous violations of the CO NAAQS. Many industrial facilities have controlled CO emissions, either directly or indirectly, in efforts to reduce VOC emissions.

Residential sources, including wood burning fireplaces, coal, oil and gas-fired space heaters, and lawn and garden engines, also produce CO emissions. Lawn and garden engines have the greatest share, approx-

---

[7]It should be noted that, since the dissociation of CO_2 to CO and O_2 is endothermic, at sufficiently high temperatures the CO to CO_2 ratio can be appreciable. High temperatures during combustion also account for the production of NO and NO_2 from N_2 and O_2 in other endothermic reactions.

imately 11 million tons per year (about half of the nonautomotive fraction). Those emissions, as well as emissions from forest fires, typically occur in summer when CO does not approach nonattainment levels. However, other sources such as chain saws and generators may be used year-round. Oxygenated fuels might reduce CO emissions from these sources.

Although space heating, particularly from wood burning, comprises a small part of the inventory (about 3%), timing and spatial scale in high-CO areas can make that contribution more significant, particularly in places like Fairbanks, Alaska; Missoula, Montana; Denver, Colorado; and other areas where wood burning for both recreational and functional purposes is common. Substituting cleaner-burning fireplaces and stoves or switching to natural gas can reduce CO and PM emissions. These controls can have a greater role in reducing human exposure in cases where the emissions are trapped in a confined space (e.g., indoors). Missoula, Montana, for example, has had a persistent problem with PM_{10},[8] and their emissions inventory for CO in 1990 attributed 28% of CO to residential wood burning (Therriault 2002). As a result, the city banned installing new wood stoves and discouraged the use of those already in operation. This reduced the contribution of residential wood burning to 18% of the CO emissions inventory by 1996.

MONITORING, MODELS, AND INVENTORIES

There are three important tools for air quality management of CO and other pollutants: monitors, models, and inventories. Monitors provide measurements of ambient pollutant concentrations. These measurements—made either on a short-term special-project basis or at permanent stations—can provide assurance that ambient concentrations do not pose a health risk to vulnerable members of society and can show where levels are high enough to potentially put an area in nonattainment of the NAAQS.

Mathematical models are used for a variety of purposes locally and regionally—from estimating current or future emissions for inventories to computing pollutant concentrations that can be expected for a given (time-

---

[8]PM_{10} is a subset of PM that includes particles with an aerodynamic equivalent diameter less than or equal to a nominal 10 micrometers.

dependent) emissions rate and set of meteorological conditions. Forecasts can be made to enable officials to announce air quality alerts or to demonstrate that proposed mitigation measures will reduce ambient concentrations sufficiently to bring an area into attainment or to maintain attainment.

Emissions inventories are assessments that identify the sources of an air pollutant in a given area and their annual contributions to total emissions. Inventories can be very helpful to policy-makers by showing which sources produce the most pollution and therefore which mitigation measures are likely to be most effective. Emissions inventories are also a useful tool in conformity determinations when compared with the emissions budget (Howitt and Moore 1999). CO emissions rates for inventories or other purposes are seldom measured directly; they are estimated using models and/or emissions factors.[9] Models are verified and improved by comparing their results with measurements of actual concentrations. Thus the three major tools of air quality management—monitoring, models, and inventories—are linked.

Monitoring

At the heart of air quality management strategies are ambient measurements. Air pollution monitoring is done at different temporal and spatial scales, depending upon the planned use of the data collected. Short-term monitoring is often used for specific projects, and long-term monitoring at permanent stations is used to determine air quality trends. In many cities, monitoring has been in place for decades, although some smaller cities were identified as CO problem areas in the 1990s and have shorter historical records. CO concentrations in most cities are falling far below exceed-

---

[9]Emission rates can be determined directly by measuring airflow rates and compositions (e.g., from a power plant smokestack or vehicle tailpipe). However, the development of emissions inventories relies on emissions factors, which express CO as the mass of an emission per unit of activity, for example, grams of CO per mile driven, or tons of CO per million kilowatt-hours (kWh) of electricity generated. The mass of CO could be calculated as the emissions factor times the annual number of miles driven or the number of kWh generated.

ance levels, so reduced support for CO monitoring is being considered (EPA 2002g).

Short-Term Monitoring

Short-term microscale monitoring is sometimes done to compile project-specific information. This type of sampling was much more prevalent in the late 1970s and the 1980s when corridor analysis for highway projects was being carried out. Now emphasis is placed on using models to predict specific, microscale pollutant concentrations. Large, controversial roadway projects may still include monitoring to assess local air quality, and research for microscale modeling still depends on measurements for model verification and validation.

The primary purpose of short-term microscale monitoring is to determine the existing concentrations of pollutants near a planned activity, such as roadway construction. Often, this type of monitoring is conducted near hot spots in the vicinity of the project. Sampling is typically done upwind and downwind near the existing source, often at multiple locations, to determine its emissions contribution. The upwind sites represent the background level (the concentration before the contribution of the source). Sampling times are related to source activity periods and are usually on the order of hours or days. Sampling may be reported at intervals as small as 1 min. Usually a very limited number of pollutants is measured, CO being the most common. Measured 1-hour and 8-hour concentration averages can be directly compared to the NAAQS.

Short-term saturation studies are especially important for understanding how meteorology, topography, and emissions activities affect the distribution of CO emissions and their effects in a location. Saturation studies attempt to develop a detailed understanding of the horizontal and vertical distribution of CO and to explore the role that physical and human factors play in elevated concentrations. Because they can also be used qualitatively in assessing the human exposure to CO, they have already been discussed in Chapter 1.

Permanent Monitoring Stations

Long-term, permanent stations are used for monitoring in most large urban areas in the United States as part of EPA's ambient air monitoring program.[10] They are often referred to as state and local air monitoring stations (SLAMS) and are quite extensive, as shown in Figure 3-5. The primary purpose of these stations is to determine attainment status and monitor air quality trends in the area, but the data serve other purposes. They are applicable in the following:

- Activating emergency control procedures that prevent or alleviate air pollution episodes.
- Observing pollution trends throughout the region, including non-urban areas.
- Providing a database for research evaluation of the effects of urban, land-use, and transportation planning; development and evaluation of abatement strategies; and development and validation of diffusion models.
- Determining highest concentrations expected to occur in the area covered by the network.
- Determining representative concentrations in areas of high population density.
- Determining the impact of significant sources or source categories on ambient pollution levels.
- Determining general background concentration levels.

Once placed, monitoring stations are only moved under special circumstances. At many of the stations multiple criteria pollutants are measured along with several meteorological variables. The data from most sites are transmitted to databases maintained by EPA. Some of the information can be accessed through the aerometric information retrieval system (AIRS) (EPA 2002h). AIRS allows long-term trends to be examined; local area average concentrations to be established; effects of abatement measures and impacts of local sources to be evaluated; and unusual occurrences to be recorded.

Whereas the results from short-term monitoring tend to be used for planning and research, data from long-term monitoring stations can have a direct impact on federal policy. Because these stations are used to determine air quality trends and to establish attainment status, the measurements made there play a large role in federal air quality management. For

---

[10]See EPA 2002g for a description of EPA's ambient air monitoring program, its objectives and proposed changes.

FIGURE 3-5 Locations of state and local air monitoring stations (SLAMS) and national air monitoring stations (NAMS). Source: EPA 2001a.

example, trends exhibited at permanent monitoring stations were part of the basis for the changes made to the national ozone standard.

Models

In recent years, much more emphasis has been placed on modeling than on monitoring to determine local concentrations. Accordingly, model development and appropriate use are crucial to the overall air quality management process. Models for mobile sources, stationary sources, and regional impacts are prescribed by EPA in Appendix W of the Code of Federal Regulations, Title 40, Part 51. States are allowed to use these preferred models to estimate local area concentrations and to compare them with the NAAQS.

Efforts to evaluate the effectiveness of CO air quality management are inherently interdisciplinary. As shown in Figure 3-6, estimating CO emis-

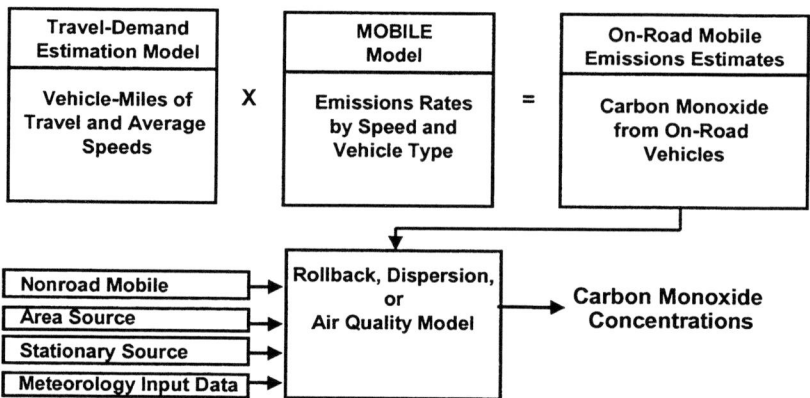

FIGURE 3-6 Use of models in the estimation of ambient CO concentrations.

sions and assessing their impacts on air quality require the interaction of three different models and the related areas of expertise: travel-demand models and other methods of estimating activity levels, emissions models, and air quality models.

Travel-Demand Forecasting

Determining emissions estimates requires data on vehicle activity, usually vehicle-miles traveled (VMT). Although direct traffic counts can be used to estimate existing emissions, travel-demand techniques are necessary to predict future traffic volumes. For existing facilities, past trends can be analyzed and extrapolated to future conditions. In the case of new facilities or future traffic volumes, demographic data is used to provide the locations of households and employment in small traffic-survey zones within the urban region and to forecast regional economic growth, land-use patterns, and future demographic trends. The change in traffic is estimated using the four-step travel-demand modeling process that encompasses.

- *Trip generation.* The estimation of the number of trips by zone per time of day and type (both trips originating in a zone, called trip production, and trips terminating in a zone, called trip attraction).

- *Trip distribution.* The pairing of trip productions with trip attractions resulting in a full spatial pattern of travel by purpose and time of day.
- *Mode choice.* The mode of travel used, specifically walk, bicycle, single occupancy vehicle, high occupancy vehicle (HOV), bus, rail, or truck.
- *Route assignment or choice.* Trips are assigned to paths in the transportation infrastructure by minimizing travel times, or travel times and costs, and incorporating average speed and other impedance feedbacks.

Travel-demand models provide the emissions factor model (typically the MOBILE model) with information on average VMT and vehicle speeds for each roadway segment that can be aggregated by roadway type or facility (e.g., freeways, arterials, collectors, and freeway ramps).

At a microscale level, traffic simulation models can be used to estimate emissions producing activities. These models assess queuing and traffic flow along specific roadway segments or at specific intersections. The models are also combined with instantaneous emissions models to predict emissions inventories (NRC 2000).

Emissions Factors Models

The emissions inventory is an important tool for estimating the relative impacts of emissions controls. Inventories predict the total mass emitted annually from all contributing sources of a particular pollutant, such as CO. Changes in sources can then be quantified in terms of the reduction in the mass released, and the control options that yield the greatest reductions can be determined. Estimated future total-mass emissions can be compared with existing emissions to determine the overall effectiveness of a mitigation program. However, as noted previously in this report, uncertainty and errors in emissions models are likely substantial. More evaluations comparing modeled emissions with observations as well as more uncertainty analysis of emissions models are needed.

On-Road Emissions Models

Because mobile sources contribute such a large fraction of CO emissions and depend on many more factors than typical stationary sources, special computer programs have been developed to estimate the appropriate emissions factors. In 49 states, the MOBILE series of models is used—MOBILE6 is the latest version (EPA 2002i).[11] California, the one exception, uses EMFAC2002.[12] Both MOBILE6 and EMFAC2002 forecast emissions factors for many different types of vehicles and fleet mixes by age, fuels, and operating conditions. Outputs of these models can be used to estimate CO emissions (in grams per mile) for starting, idling, and running.

Uncertainties and errors in MOBILE are well documented (e.g., Hallock-Waters et al. 1999; NRC 2000; Sawyer et al. 2000; Holmes and Russell 2001; Frey and Zheng 2002; Parrish et al. 2002; Pokharel et al. 2002). Concerns over these issues prompted a recent NRC committee charged with reviewing MOBILE to recommend that enhanced model evaluation studies, in tandem with uncertainty studies, begin immediately and continue throughout the long-term evolution and development of mobile-source emissions models (NRC 2000; Holmes and Russell 2001). The committee recommended that such studies be done with oversight and guidance from an outside reviewing body that includes users and technical experts.

MOBILE6 contains many improvements on previous versions of the model. However, Dulla and Heirigs (2002) expressed concern that MOBILE6 understates the benefits of technology improvements on CO emission rates, especially in areas with colder climates such as Alaska where the benefits of off-cycle controls are assumed to be negligible during winter driving. The modest decline in CO emissions rates forecast by MOBILE6 is much lower than what would be expected given the introduction of control-system technologies designed to meet NLEV and Tier 2 requirements. One reason for the modest projection is that forecasts of CO emissions rates are extremely sensitive to gasoline sulfur content levels when model runs use the standard assumptions for Fairbanks (including no benefits from the SFTP). Because MOBILE6 forecasts for CO emissions in Alaska are largely a function of projected gasoline sulfur levels and have little to do with improvements in emissions control designs, Dulla and Heirigs (2002) are concerned that projected CO emissions rates understate the benefits of NLEV and Tier 2 technologies and may overstate the level of local controls needed.

---

[11]See EPA 2002j for a detailed description of the MOBILE model.
[12]See CARB 2002 for a description of the EMFAC model.

Nonroad Sources

Airports are important nonroad sources of CO emissions. To help in the preparation of airport emissions inventories, the Federal Aviation Administration has released the Emissions Dispersion Modeling System (EDMS), Version 4.[13] This computer model includes emissions rates for most aircraft types, ground support equipment, motor vehicles (based on MOBILE), typical stationary sources at airports, and training fires. The model requires operational data for each of these sources and generates an emissions inventory for the airport. EDMS contains a Gaussian dispersion model, which is discussed in the following section, and also requires meteorological data to predict local CO concentrations.

Construction equipment contributes a large amount of nonroad emissions. Until recently, these sources were mainly ignored because their emissions were thought to be of short duration. Large multiyear projects, such as the "Big Dig" in Boston (Big Dig 2000), have greatly changed this outlook. Studies have shown that construction equipment can exacerbate local air quality problems, and off-road sources are now considered more often than they were before. The contributions of other nonroad engines and off-road vehicles listed in Table 1-2, such as off-road recreational vehicles (snowmobiles, in particular), lawn and garden equipment, and boats, have also come under review as additional CO reductions are sought. In some locations, their emissions can be substantial.

As emissions from on-road vehicles decrease because of tighter emissions standard, fuel-sulfur controls, and less deterioration of emissions control devices, the emissions estimation techniques for nonroad mobile sources will continue to increase in importance. The NRC (2000) concluded that, primarily because of a lack of data, the current off-road emissions model (NONROAD) does not accurately estimate emissions from off-road vehicles and nonroad emissions inventories or the effects of emissions controls on these sources. That report recommended a major effort be undertaken by EPA to update NONROAD with better population and activity data and real-world emissions factors. EPA has been developing enhancements to the NONROAD model to improve emissions estimates.[14]

---

[13] See FAA 2002 for a description of the EDMS.
[14] See EPA 2002k for a description of NONROAD.

Stationary Sources

Stationary sources, especially power plants and large industries, may also have a large impact on local CO concentrations. As previously mentioned, stationary-source emissions factors can be determined from AP-42 (EPA 1995). Stationary-source operations are usually more consistent than mobile-source operations, thus stationary emissions are easier to quantify. Activity, such as fuel usage (often in the form of BTUs generated or amount of fuel consumed per year), is multiplied by an emissions factor to estimate the total mass of CO emitted per year. However, CO exceedances in Birmingham, Alabama, demonstrate that an unregulated point source that experiences process upsets can become a large emissions source sufficient in itself to create CO exceedances. Utilizing emissions factors from AP-42 would underestimate the contribution from sources such as the one in Birmingham.

In addition, estimating emissions from area sources, such as residential heating, is likely to be highly uncertain. During this study, the committee noted that the emissions inventory for Missoula, Montana, attributed 18% of CO emissions to wood burning, whereas the inventory for Fairbanks, Alaska, attributed only 3% of CO emissions to that source. The disparity existed despite Missoula's fairly substantial effort to control emissions from wood stoves. The committee also questioned whether the increasing popularity of fuel-oil stoves has resulted in the underestimation of this source in inventories. It is clear that emissions inventories for stationary sources need improvement.

Air Quality Models

Air quality modeling is an essential element of air quality management. Models can be used to evaluate plans for attainment of an NAAQS (also referred to as an attainment demonstration), to evaluate the effects of new construction projects, and to conduct further research into what causes pollution episodes and how they can be predicted. A number of modeling techniques—requiring various levels of scientific expertise, input data, and computing resources—are available for these purposes. The simplest models, rollback models, assume a direct correlation between emissions and ambient pollutant concentrations; the most complicated models, grid-based air quality models, resolve temporal and spatial variations in pollutant concentrations and the effects of meteorology, emissions, chemistry, and topography. Models are also characterized by the size of the problem they address: microscale models simulate pollution from a point source or intersection; mesoscale models simulate metropolitan or multistate pollution; and large-scale models simulate continental or global pollution.

In attainment demonstrations presented in SIPs, states are required by EPA to model how emissions reductions will lead to the desired air quality improvements. Three types of models have been used to demonstrate attainment of the CO NAAQS: rollback (also knows as statistical rollback), Gaussian dispersion, and numerical predictive models.

Rollback Models

The simplest of the three models used for attainment demonstrations is the statistical rollback model in which the needed reduction in emissions is assumed to be proportional to the required reduction in ambient CO concentrations (ADEC 2001).

$$\% \ reduction = \frac{CO_{baseyear} - CO_{NAAQS}}{CO_{baseyear} - CO_{background}},$$

where

$CO_{base\ year}$ = the second highest 8-hour average in the base year;
CO_{NAAQS} = the NAAQS of 9 ppm (or sometimes 9.4 ppm); and
$CO_{background}$ = an average regional background CO in the absence of emissions.

Although easy to implement, rollback models do not explicitly consider the role of meteorology or the spatial heterogeneity of CO emissions and concentrations. EPA has allowed states to use rollback models rather than the more resource-intensive dispersion and urban-airshed models described below, to demonstrate attainment in smaller cities. An improvement on the simple rollback model is the probabilistic rollback model used in CO modeling for the Puget Sound area of the State of Washington (Joy et al. 1995).

Gaussian Dispersion Models

A second type of model that has been used for CO-attainment demonstrations is a Gaussian dispersion model, which is typically used to simulate CO concentrations for microscale analysis in the vicinity of intersections or along major traffic corridors (EPA 1992). One of the first effective Gaussian dispersion models for mobile sources was CALINE3, which is still in use. Inputs for this model include meteorological data, such as windspeed and atmospheric inversion strength in the vicinity of the pollutant source, and temporally resolved emissions. Emission factors developed from other emissions models (MOBILE and EMFAC), along with traffic volumes, roadway geometries, and intersection information, are used to determine the emissions along a roadway. Dispersion modeling then includes transport and mixing to calculate local concentrations. The model is Gaussian in nature, meaning it assumes that a plume of pollutant gas released from a point source can be described by a widening Gaussian function (a bell-shaped curve) as it travels downwind (Wayson 1999). The model also makes the assumption that roadway segments can be cut into small sections with a point source approximation applied to each and their plume concentration contributions summed at a receptor site. This concept allows roadway curves or winds nearly parallel to the roadway to be modeled effectively.

The shortcoming of CALINE3 is that it is only useful for vehicles that are moving at a constant rate of speed. At locations of high CO emissions (such as intersections), increased emissions due to vehicle delay and idling must be accounted for. To do that, two models are in use today: CAL3QHC and CALINE4. Both use the same general approach to estimate dispersion as CALINE3 does. CAL3QHC is used in 49 states, and CALINE4 is used in California.

Gaussian dispersion models are typically used for local area (microscale) analysis and are used extensively in CO-related evaluations, including project-level conformity determinations. Modeling is done for the worst hour to compare with the 1-hour average CO NAAQS. Worst-case conditions (a windspeed of 1 MPH and a stable atmosphere) are often used. A persistence factor, which is a multiplier of the peak 1-hour concentration that is based on changes in wind patterns and traffic, is used to estimate an 8-hour average concentration for comparison with the 8-hour NAAQS. The model results often determine whether a project can go forward.

The American Meteorological Society (AMS) policy statement on dispersion modeling (Hanna 1978) concluded that these models are accurate within a factor of 2 for reasonably steady horizontally homogeneous conditions; however, they will be less accurate, for example, when obstacle wakes flows (e.g., from buildings or vehicles) and extremely stable thermodynamic lapse rates occur. Dispersion accuracy will also be lower, as listed in the AMS statement, for "dispersion over forests, cities, water and rough terrain."

Grid-based Air Quality Models

The most complicated models used for attainment demonstrations simulate how a pollutant concentration varies with time and space over an entire urban area. These numerical predictive models, generally intended for regional analysis, can simulate emissions from multiple sources and the dispersion, advection, and photochemical reactions of gaseous pollutants in the atmosphere. These models are integrated separately from meteorological models. Grid-based models, such as Models-3 and the urban airshed model (UAM), have been used for many years to simulate O_3, which is a region-wide or mesoscale pollutant. The UAM has been adapted to simulate CO in Denver (Colorado Department of Public Health and Environment 2000). Because of the local nature of high-CO episodes, extensive modeling of the entire urban airshed may be unnecessary for CO-attainment demonstrations. Airshed modeling is resource-intensive, requiring detailed knowledge of an area's meteorology (usually based on the output of a mesoscale weather model constrained by observations), spatially and temporally resolved emissions inventories, and measurements of the pollutant at several locations to allow model evaluation. Highly trained personnel are needed to conduct the simulations.

More complicated models are not always appropriate for attainment demonstrations, but they can be valuable in improving the understanding of the interactions among atmospheric processes. Even better research tools than the numerical predictive models described above (such as Models-3 and the UAM) are process numerical models, which allow processes specific to air quality modeling and meteorology to be coupled within a single computational framework. Process numerical models typically are formulated by adding pollutant emissions, chemistry, and transport into an existing meteorological model rather than simply using the meteorological data as a model input. The relatively nonreactive behavior of CO makes it an ideal chemical species for simulation in a weather model. Predictions of CO, for example, can be straightforward in the National Weather Service Eta model,[15] which has a horizontal grid framework

---

[15] See NWS 2002 for information on the NWS Eta model.

of 12 × 12 km over the contiguous United States. However, this resolution is insufficient for most CO problem areas. Initial work to simultaneously simulate atmospheric flow and diffusion of CO at high spatial and temporal resolution is described by Fullerton (2002).

Box Models

Box models are another tool available for microscale analysis of air pollution. The "box" is some volume of air into which emissions are injected. Box models may divide a region into cells of equal volume and use mass balances to treat the transfer of CO among cells. In their simplest application, they can consist of a single box. The cells may also be separated in the vertical direction. Air within each cell is assumed to be well mixed. Simplifications of this concept lead to the common single-cell box model.

Though box models are not used in attainment demonstrations, they are particularly useful to understand how various emissions scenarios and meteorological conditions affect pollutant concentrations. For example, a box model for CO in Anchorage, Alaska, has been used to quantify how mechanical turbulence from roadway traffic might increase the mixing height and reduce CO concentrations on severe-stagnation days compared with concentrations observed in residential neighborhoods (Morris 2001). Appendix C describes a single-cell box model, with and without recirculation. The committee's interim report on Fairbanks describes the application of such a model to Fairbanks, Alaska (NRC 2002).

Summary of Air Quality Models

There is no single air quality model that is the best for CO for all locations. Typically the choice depends on the severity of the problem, the available data, and the resources available for modeling. It its interim report (NRC 2002), the committee recommended that Fairbanks, Alaska, use a simple box-model approach for air quality planning purposes in the near term. A box model simulates the effects of emissions and meteorology in a well-mixed controlled volume. The committee felt that such an approach could provide greater insights into the effects of the timing of CO emissions and of meteorological variables, in this particular situation, given the limited vertical dispersion and available data. Box models could supplement Fairbanks's current approach of using a simple rollback model, which they used in their attainment demonstration (ADEC 2001).

In the long-term, the committee recommended that more work be done to develop, apply, and evaluate more sophisticated, physically comprehensive models that would simulate how CO concentrations vary with time and space. Because CO is relatively conservative on time scales of hours, knowledge of the temporal and spatial distribution of CO emissions and of the observed CO concentration field provide an effective diagnosis of atmospheric dispersion patterns. For chemical species that are eliminated by reactions in the atmosphere, knowledge of the CO dispersion provides an observational constraint on the concentration fields of the reactive species.

The committee concluded that more physically comprehensive models should be used for planning, forecasting, and assessing human exposures to high CO concentrations. It is important that model development and testing be specific to the extreme conditions that occur in CO problem areas such as Fairbanks. However, model development must occur in concert with improved monitoring to enable model evaluation. The committee believes that even in areas such as Fairbanks, which has experienced very few exceedances since 1996, and none since 2001, the development of comprehensive models is still worthwhile. The number of periods of elevated CO levels experienced in Fairbanks indicates that the city is still susceptible to exceedances. Furthermore, CO modeling can be used to better understand and characterize CO hot spots as well as other criteria pollutants and air toxics. The development of a better modeling approach today will benefit all problem areas in the future.

Despite advances in air quality modeling capabilities over the last 30 years, many improvements are still possible and necessary. One problem is that the vertical and horizontal resolution of models is too coarse to capture the variability in pollutant concentrations, which is necessary to identify local hot spots and is important for determining local concentrations downwind of hot spots. In addition, the validity of model representation becomes questionable when unusual meteorological conditions occur, and that could lead to errors in the prediction (Pielke 2002). Models used for regulatory purposes can suffer a loss of realism as a result of such shortcomings, leading to costly errors in planning. Models also need more realistic three-dimensional dynamics (advection, pressure gradient forcing, turbulence) and more realistic parameterizations of smaller-scale processes (e.g., turbulence from buildings, radiative flux divergence changes in the temperature profile associated with aerosols in the lower levels of the atmosphere). The models also need higher spatial and temporal resolution. Ensemble runs of the models should be performed to provide a more realistic envelope of simulated dispersion patterns. However, the committee recognizes that this adds cost and time to the evaluation. Not only can these models be used for air quality applications, models with higher resolution can also assist in homeland defense because they can help understand the dispersion of accidental or deliberate releases of chemical, biological, and radiological materials.

In 2003, a large dispersion research project will be undertaken to help define important dispersion parameters, primarily for homeland security purposes (DOE/DOD 2002). The project will be a month-long study conducted by a combination of federal and state governmental agencies with support from multiple universities. Research will include releases of tracer gases with careful measurements of meteorological parameters to determine dispersion trends for city-wide dispersion, dispersion in street canyons, infiltration to buildings, and effects of topography.

Statistically Robust Methods to Assist in Tracking Progress

The air quality models described above assess the effectiveness of emissions controls and the prospects for attaining the CO standard by representing critical processes within a physically based model of the system. An alternative to those physical models is to take a statistical approach assessing the relationship among human activities, CO emissions, meteorology, and ambient air quality, as described below.

Probability of an Exceedance

Reddy (2000) carried out an analysis of the probability of a future CO exceedance in Denver that might be broadly applicable to other areas. The analysis took into account the historical variability in CO concentrations as a result of meteorology and unusual traffic events. The purpose of his analysis was to determine the risk of a CO exceedance associated with eliminating or altering the oxyfuels program during the first week in February for the future years 2002-2013. He used CO monitoring data from the CAMP site (AIRS ID 08-031-002), which is the site in Denver that has historically shown the greatest number of exceedances. He used daily peak

8-hour average CO concentrations for the first week in February for the 20-year period of 1975-1994. Because these values depended on the emissions during those years in addition to stochastic meteorology and occasional unusual traffic, Reddy corrected past CO concentrations for each year to what they would have been if the emissions for that year had been the same as those projected for 2002.

The natural logarithms of the corrected peak 8-hour average CO concentrations were normally distributed; the 8-hour averages themselves were not. By estimating future emissions inventories for the years 2002-2013, based on projected fleet composition and VMT, and assuming that the lognormal distribution would hold for future years, Reddy was able to calculate the probability of an exceedance on a single day of the first week in February (P_{1-d}) for the future years. He then used Equation 1 to compute the probability of one or more exceedance days during an entire week (P_{7-d}).

$$P_{7-d} = 1 - (1 - P_{1-d})^7. \qquad (1)$$

Reddy found a better than 5% chance that an exceedance might occur if Denver immediately suspended the oxyfuels program for the period 2002-2013. The study also found that Denver would likely not have an exceedance if 1.5% oxygen (which is less than the oxygen content used in the current oxyfuels program) was used in fuels for 2002 and 2003 before suspending the use of oxyfuels for 2004 through 2013.

Equation 1 assumes that exceedance events are independent over time (thus the probabilities can be multiplied, as in the second term on the right hand side of the equation). The assumption might not hold; for example, exceedance events might be positively associated over time. Given this possibility, Reddy's method might overestimate P_{7-d}. Alternatively, we can modify Reddy's equation as follows:

$$\text{Expected number of exceedances} = N \times P_{1-d}, \qquad (2)$$

where N denotes the number of days in the time period being considered, under the assumption that the probability of exceedance is uniform over the time period. For Reddy's application, the time period considered is the first week in February, thus $N = 7$. Under more general conditions, Equation 2 can be modified as follows:

$$\text{Expected number of exceedances} = \sum_{i=1}^{N} P_i, \qquad (3)$$

where P_i denotes the probability of exceedance on the i-th day. Equation 3 does not assume that the exceedance probability is uniform over time. For example, one might use a different exceedance probability for weekdays versus weekends.

The same procedure that Reddy used, or the modified one discussed above, could be applied to monitoring sites in other cities and for times other than the first week in February (e.g., a whole winter season), provided that there are enough historical data to establish the distribution of CO concentrations and to estimate emissions inventories for past and future years.

Meteorological De-trending

Ambient CO concentrations across the nation are going down. Undoubtedly many of these reductions are due to emissions controls. Part of the trend, however, may also be meteorological. A warmer winter with less stagnation can lead to lower winter CO levels. As noted by Neff (2001), Denver may be experiencing lower CO levels than would be expected from emissions reductions alone because of warmer winters with greater vertical mixing. How can the impact of meteorological trends on the observed concentrations be removed in order to assess the impact of emissions controls and to show true progress towards meeting air quality standards in the future, when meteorological conditions may not be so favorable? One must "de-trend" the observations.

Meteorological de-trending is accomplished by identifying how meteorological variables impact pollutant concentrations and removing the influence of those variables. One way would be to create a physically realistic model that can simulate many years, developing emissions-to-air quality relationships and showing how they respond to meteorological influences. However, this approach would be cumbersome and would introduce significant uncertainties. The influence of meteorology is more typically identified using an empirical approach. Many years worth of concentration data are analyzed, along with the corresponding meteorological data, to develop a statistically based model. That model is then used to remove meteorological impacts (Kuebler et al. 2001; Porter et al. 2001).

> **Recommendations: Ambient CO Modeling**
>
> More sophisticated, physically comprehensive models that can simulate how CO concentrations vary in time and space should be developed, applied, and evaluated. Ongoing research should be continued. Such models would be used for air quality planning and forcasting and for assessing human exposure to high concentrations of CO and related pollutants. Because CO is a relatively unreactive pollutant, the ability to better represent CO's temporal and spatial distribution provides an effective diagnosis of atmospheric dispersion patterns. Model improvements would ahve applications for other air quality management issues and would offer the potential to better understand the dispersion of chemical, biological, and radiological materials. Most importantly, improved models will permit more effective and realistic planning, leading to better-informed decisions by administrators. Model development should occur in concert with improved monitoring to enable model evaluation. In addition, the statistical forecasting models should be improved.

Recent work by Flaum and colleagues used a multistep process to resolve the trends in ozone (O_3) into four components: a long-term trend, presumably due to emissions controls; a seasonal component; a component driven by meteorological fluctuations; and a noise component (Flaum et al. 1996). Kuebler et al. (2001) used a similar approach, not only for O_3, but also for CO, NO_x, and VOCs, and compared the meteorologically detrended concentrations of the primary pollutants with the trends in emissions estimates. From that, a direct relationship between the emissions levels and pollutant concentrations could be established.

The latter approach appears appropriate here given its prior use for CO, though the explanatory variables may depend on location. For example, in Fairbanks, a nonlinear response to temperature is expected because CO concentrations appear to be highest at about -20°F to 20°F, not at much lower or much higher temperatures. This approach is convenient for local air quality management organizations because it requires relatively little data (e.g., a long-term record of CO concentrations and meteorological variables such as temperature and windspeed would suffice, though more factors are useful) and nominal computational power.

The de-trending analyses also can provide extra information for air quality planning. As noted above, de-trending can be used to help develop probabilities of exceeding the NAAQS for CO at various emissions levels. From that, the necessary level of emissions can be identified in a more statistically robust fashion.

4

The Future of Carbon Monoxide Air Quality Management

When regulations on carbon monoxide (CO) automobile emissions began in the 1960s, large areas within many cities were experiencing high CO concentrations. Motor-vehicle emissions controls over the past three decades have greatly reduced ambient CO concentrations. As shown in Chapter 1, the number of monitors showing CO violations has fallen to only a few, and the monitors that still show violations do so much less frequently. CO control has been one of the greatest success stories in air-pollution control. As a result, the focus of United States air quality management has shifted to characterizing and controlling other pollutants, such as tropospheric ozone, fine particulate matter ($PM_{2.5}$),[1] and air toxics.

However, as described in Chapter 2, some locations will continue to be susceptible to violations of the CO health standard because of meteorological and topographical conditions that produce severe winter inversions. In addition, CO remains relevant to air quality managers because it acts as an indicator of a host of other mobile-source-related emissions, and its control produces substantial collateral benefits. Understanding the distribution and effects of CO exposure will be a continuing challenge. The fixed-site com-

---

[1]$PM_{2.5}$ is a subset of particulate matter that includes those particles with an aerodynamic equivalent diameter less than or equal to a nominal 2.5 micrometers (μm).

pliance monitoring system, although important for understanding overall emissions and air quality trends and area-wide compliance with standards, is incapable of capturing all locations that exceed the standards and fully characterizing the spatial variability of CO. The recent exceedances in Birmingham, Alabama, indicate that there may be locations that have frequent, unmonitored exceedances. Personal exposures to CO and related pollutants are also not represented by these fixed-site monitors. In this concluding chapter, the committee focuses on the issues related to exposures and CO management that will be important in the future.

EXPOSURES OF CONCERN IN THE FUTURE

Proximity to CO emissions sources determines human exposure profiles, blood levels of carboxyhemoglobin, and risk of adverse health effects. In short, place makes the poison (Smith 2002). Personal exposures may vary by individual depending on occupation and personal habits. Individuals who have long commutes or who drive for a living can be exposed to CO levels well in excess of those measured at fixed-site monitors. Operation of nonroad sources, such as construction equipment, gasoline-powered lawn and garden equipment, snow blowers, snow machines, and other recreational vehicles, may result in significant personal exposures. Maladjusted home heating units operated in confined spaces and unventilated homes remain sources for high CO exposures, as does cigarette smoking. These sources expose individuals to high concentrations but have no measurable effect on any fixed-site monitor.

In addition, CO concentrations are not uniform across a region, and hot spots with higher levels of ambient CO may occur at discrete locations. Hot spots often occur in places with high vehicle traffic or other local sources, especially when topographical and meteorological conditions are conducive to CO accumulation. Demographic data (see Table 1-8 in Chapter 1) also indicate that the monitors that have recently shown violations of the health-based CO NAAQS tend to be in low-income urban areas. Because CO is a good indicator of exposure to other air toxics generated by mobile sources, hot spots may identify locations where individuals are at higher risk for adverse health effects from a number of urban air pollutants.

The committee recommends an active program to identify and characterize hot spots and to better define the upper end of the CO exposure distribution. Current CO monitoring technologies are sufficiently advanced that accurate mobile monitoring can be deployed in major urban areas and can be combined with temporary fixed-site monitors to produce the data necessary to simulate actual exposures experienced by the general population. Monitor-equipped vehicles can traverse randomly sampled routes and destinations in the region, sampling CO concentrations as they move through the road system. A national program employing mobile, temporary fixed, and possibly personal monitors would provide tremendous benefits for exposure assessment, health impacts analysis, model evaluation, and attainment and maintenance planning.

FUTURE CO MANAGEMENT ISSUES

Future management of CO will contend with both the changing nature of the CO problem and the changing nature of the air quality management system. This section discusses the roles of new-vehicle emissions standards, oxygenated fuels, and transportation-control measures in managing CO; the appropriate spatial scale for CO management; and the possibility that increasing VMT and other factors may counter the decline in vehicle CO emissions per mile. The chapter concludes with a section discussing the integration of CO into the overall management of air quality in the United States.

Improvements in Vehicle CO Emissions

Emissions from light-duty trucks and passenger cars will continue to be the focus of CO management in the future. Current CO controls of importance include Tier 1, NLEV, cold-temperature standards, the SFTP, and I/M and/or OBDII. Of greatest importance for future CO emissions control are the effectiveness of cold-temperature, Tier 2 vehicle emissions standards and the use of low-sulfur gasoline under meteorological conditions conducive to CO buildup.

The cold-temperature CO standards have provided significant reductions in emissions during the first few minutes of engine operation at low temperatures. For northern cities such as Fairbanks, Alaska, a more stringent CO cold-start regulation (or a lower cold-start test temperature) would be beneficial for further reducing emissions under conditions that favor exceedances. The committee discussed this option in its interim report on

> **Recommendations: Federal Tier 2 and Cold-Start Emissions Standards**
>
> In the absence of compelling evidence, the committee does not recommend tightening the national cold-start standard below 10.0 g/mi or requiring that the 10.0 g/mi standard be met at a lower temperature. However, supplemental emissions testing should be undertaken at temperatures below 20°F to determine to what extent CO emissions systematically increase as ambient temperature decreases. Testing data should be obtained and analyzed at 0°F and 10°F, and should include CO as well as other pollutants (air toxics and PM).
>
> The extent of the anticipated reduction in CO emissions from Tier 2 vehicles needs to be confirmed through analysis of data including cold-start data at 0°F and 10°F. Again, testing should include CO as well as other pollutants. If the analysis of Tier 2 and prior controls indicate that all locations will attain the 8-hour CO standard, more stringent federal CO vehicle-emissions standards will be unnecessary. The results of all emissions testing must be incorporated into EPA's MOBILE model to accurately estimate future CO emissions. The effects on CO problem areas of the sales strategy used by manufacturers to meet the NO_x limits as well as the trading and banking provisions also needs to be assessed and incorporated into emissions modeling.

Fairbanks and concluded that toughening standards should be considered in other geographical areas that might also benefit. In Chapter 3 of this report, the committee reviewed testing information on cold starts and on new emissions control technologies.

In the absence of compelling evidence, the committee cannot recommend making the cold-start CO standards more strict. In the future, additional fleet-average CO emissions reductions will come from the increased number of vehicles certified to cold-temperature standards and from the introduction of emissions control technologies and low-sulfur fuels that will be adopted to meet the Tier 2 standards. However, supplemental testing should be done to assess emissions performance below 20°F. Testing should also be done to determine whether existing onboard diagnostic (OBD) systems operate properly at 20°F. In addition, CO emissions reductions from Tier 2 vehicles must be confirmed, especially during cold starts below 20°F.

> Recommendations: Oxygenated Fuels
>
> The EPA should undertake a science and policy review of the current oxygenated fuels programs to determine the conditions under which these programs are cost-effective. The review should also determine when changes in fleet technologies will render these programs ineffective. Low-temperature testing, especially below 20°F, is recommended. Oxygenated fuels programs should be implemented only when they provide cost-effective reductions in CO that help areas come into compliance or prevent areas that have attained the NAAQS from falling back into nonattainment.

Oxygenated Fuels Program

The use of oxygenated fuels (or oxyfuels) is required in all areas of the United States that exceed the NAAQS for CO. Oxygenated fuels programs have declined in effectiveness and are expected to continue to decline as more modern vehicles enter the fleet.[2] Therefore, the question arises: Should a mandatory oxygenated fuels program continue?

An oxygenated fuels program aimed at reducing winter CO emissions appears to be of decreasing value. However, malfunctioning vehicle emissions systems, which might benefit from use of oxygenated fuels, could dominate vehicle emissions in the future. The committee concludes that EPA should undertake a review of the science and policy behind the current oxygenated fuels programs to determine the conditions under which these programs are cost-effective. Low-temperature testing, especially below 20°F, is recommended. The review should also determine when these programs will no longer be cost-effective because of changes in fleet technology.

Transportation Control Measures for Reducing CO Emissions

Further reductions in emissions may be aided in the future by transportation control measures (TCMs). TCMs seek to reduce tailpipe emissions

---

[2]An oxygenated fuel is a gasoline containing an oxygenate, typically methyl *tertiary*-butyl ether (MTBE) or ethanol, that is intended to reduce production of CO.

per mile through improvements to traffic flow and to reduce vehicle-miles traveled (VMT) through the management of transportation demand. However, TCMs have accounted for only a small share of the overall reductions in emissions. Studies show that traffic-signal coordination and control strategies can reduce fuel use from 8% to 15% in specific corridors. These strategies would presumably reduce CO hot spots as well. However, total regional impacts of these control strategies might only be a 1% to 4% reduction in fuel use (Cambridge Systematics, Inc. 2001). Although the empirical evidence is limited, the most successful efforts to manage transportation demand have probably resulted in reductions in VMT of considerably less than 1% (Cambridge Systematics, Inc. 2001). Many TCMs, particularly pricing strategies, have proved to be unpopular and politically infeasible, and those that have been easy to implement, such as voluntary trip-reduction programs, have shown limited effectiveness. On the bright side, transportation agencies are increasingly experimenting with new TCMs, some of which are listed in Table 3-5. The challenge for these agencies in the future will be to adapt and combine TCMs effectively to meet the specific needs of their region.

Spatial Scale for CO Management

One critical issue is determining the most appropriate spatial scale for CO management—national, regional, or local. Although national-scale controls, especially vehicle emissions certification standards, have played and will play a crucial role in eliminating most exceedances of the CO standards, future CO attainment will continue to depend on regional and local control strategies. However, the design of appropriate strategies depends on a more thorough understanding of the sources of CO emissions in specific areas and the meteorological, topographical, and human factors that contribute to the formation of hot spots.

The current process of using regional CO emissions inventories for the analysis of localized exceedances demonstrates this issue. When regional inventories are used, some sources that do not contribute to local exceedances might be included in the control plan. In the case of Fairbanks, the committee noted that the state implementation plan (SIP) emissions inventory contains some sources elevated well above the atmospheric inversion height (which may be as low as 10-20 feet) as well as some distant sources (over 10 miles away). In Fairbanks, the same meteorological conditions that trap automotive exhaust near the ground keep emissions from elevated sources from reaching the ground. Determining what sources actually contribute to an exceedance at a site would require wind pattern and/or tracer studies. Given the nature of the meteorology in Fairbanks, the committee feels that it is unlikely that all of the sources used in the SIP actually contribute to CO exceedances. Such difficulties illustrate the need to improve the spatial and temporal resolution of emissions and meteorological variables through improved monitoring and modeling in locations that still exceed the CO standard.

A better understanding of the factors that contribute to localized CO problems can provide a basis for the development of more effective control strategies. For example, CO exceedances at the Lynwood, California, monitor are attributed to a combination of high vehicle emissions and local topographical and meteorological conditions. Little is known, however, about the major sources of traffic moving through the area. Strategies to address emissions from through-traffic might include regional controls as well as local improvements to traffic-signal controls. Strategies to address emissions from locally owned vehicles might include vehicle buy-back or repair assistance programs targeted to eliminate high emitters.

The committee advises against focusing solely on local controls that only affect sources in a small geographical area. Given the inability of the fixed monitoring stations to represent the full spectrum of exposures and the mobility of the largest source of CO (the automobile), such an approach would not ensure that the wider region is adequately protected against CO and related pollutants. The elimination of hot spots will undoubtedly require region-wide as well as location-specific efforts.

Impact of Increasing VMT and Longer Fleet Turnover

Advances in motor-vehicle emissions control technology have reduced CO emissions faster than VMT has increased. Vehicle emissions of CO are an order of magnitude less now than they were in 1970 for the same VMT. However, as VMT increases, emissions reductions resulting from control equipment might not be enough to compensate, and overall CO emissions may begin to increase at some point in the future. The increased durability of vehicles may also play a role; owners are keeping their vehicles longer. The result is that a smaller percentage of lower-emitting vehicles are entering the fleet each year, and the average fleet age is increasing. Present-day vehicles show a deterioration in the effectiveness of emissions controls over time, which is also likely in future vehicles.

An aging fleet and increased VMT could result in an exacerbation of the CO problem in existing nonattainment areas and an increased risk of a return to nonattainment status in maintenance areas. As described in Chapter 3, Davis (2001) documents the increase in the national average age and median lifetime of in-use passenger cars. The Energy Information Agency (EIA 2003) forecasts an increase in VMT by LDVs by 2.4% per year through 2020, which is larger than the 2.2% growth projected just 1 year earlier (EIA 2001). Some agencies have forecast that on-road CO emissions will begin to rise again after about 2005 due to increasing VMT (New York State Department of Environmental Conservation 1999; Colorado Department of Public Health 2000; City of Fort Collins 2001). However, those estimates are based on earlier versions of the MOBILE model. The same result will not hold when emissions projections are updated using MOBILE6, because MOBILE5 overestimated the increase in CO as vehicles age (the deterioration rate). Albu (2002), in contrast, forecasted fleet turnover would continue to decrease mobile-source CO emissions in California through 2020. CO problem areas with a greater percentage of older vehicles or where VMT is rapidly increasing because of population growth or other behavioral changes (localized increases in traffic) are expected to be the most susceptible to exceedances.

INTEGRATING CO CONTROL INTO THE OVERALL AIR QUALITY MANAGEMENT SYSTEM

Historically, all criteria pollutants have been considered independently. As such, CO tends to be managed in isolation even though other pollutants have similar emissions sources and CO can play a substantial role in the formation of ozone (O_3) (NRC 1999). Because hydrocarbon emissions have dropped over the past few decades, CO is causing a more significant fraction of the tropospheric O_3 in urban areas; yet CO emissions reductions are typically not pursued for the control of urban O_3. Further, CO control can reduce emissions and resulting exposures from other mobile-source pollutants.

The ability to use similar assessment tools and controls for copollutants can provide savings for areas that have related problems. For example, areas facing both CO and $PM_{2.5}$ problems will use regional $PM_{2.5}$ modeling

> Recommendations: Management of CO
>
> Communities with special CO problems should be encouraged to design locally effective programs. Federal and state assistance should be provided to these communities for characterization and implementation of management options. This should include assistance to improve non-motor-vehicle emissions characterization. Because the CO standard is health-based, all communities need to be diligent in working toward attaining and maintaining the CO standard. In addition, the utility of programs implemented to reduce CO emissions should be reassessed periodically. This reassessment should include their impact on CO as well as other pollutants and their impact at low temperatures.
>
> CO management should be better integrated into air quality management. Although the focus of air quality management in the near future will be on other air pollution issues, winter inversion conditions not only affect CO buildup but can also be related to higher concentrations of $PM_{2.5}$ and some air toxics. In addition, the primary source of CO, fuel-rich operations of light-duty vehicles, is a major source of other pollutants of concern. The committee therefore recommends that EPA assess the relationship of CO to these other pollutants when the CO criteria are updated.

to account for the impacts of controls and regional growth on $PM_{2.5}$ levels. Including CO in such modeling requires few additional resources and could help identify hot spots for other pollutants. In addition, the primary source of CO—fuel-rich operations of light-duty vehicles—is a major source of other pollutants of concern, including $PM_{2.5}$ and air toxics. There are many control programs that provide benefits for all of these pollutants, including those aimed at reducing cold-start emissions, removing or repairing high-emitting vehicles, and improving the effectiveness of the vehicle catalyst (e.g., with low-sulfur gasoline).

Thus, the committee recommends that CO be better integrated into the management of other related pollutants. The committee recognizes that the focus of air quality management in the near future will be on attaining the new $PM_{2.5}$ and 8-hour O_3 standards as well as reducing air toxics. However, winter inversion conditions that characterize the remaining high-CO areas not only affect the build up of CO but they are also related to higher concentrations of $PM_{2.5}$ and some air toxics. The committee recommends that EPA assess the relationship of CO to these other pollutants of concern when the CO criteria are updated.

> Recommendations: Permanent Monitoring
>
> Because of the value of this information for air quality management in general, agencies should resist removing CO monitors in locations not expected to show violations. Instead, they should consider continuing operations at existing CO monitoring sites, noting that when monitors are colocated the incremental costs of continued operation may be relatively small compared with the data's usefulness for purposes beyond demonstrating attainment. Permanent monitors also need to reflect changes in growth and development patterns to accurately assess the local air pollution situation. However, communities that have attained the CO standard with an adequate level of protection of safety might not be willing to pay to obtain data from these monitors. Support from federal and other sources might be necessary to continue monitoring operations.

Finally, the issue of integrated air quality management is important in considering the reallocation of CO monitoring resources. Very few areas in the country continue to violate CO standards. Because CO nonattainment is very unlikely in many regions, state and regional air quality planning agencies have expressed interest in reducing or eliminating CO monitoring in their areas. From a scientific standpoint, current CO monitors provide valuable information for long-term air quality management planning. The marginal savings of eliminating the remaining CO monitors, especially those co-located with monitors for other pollutants, should be weighed against the continuing benefits of collecting these CO data for use in planning, analysis, and assessment. These data can be very useful in illustrating long-term trends, evaluating the regional effectiveness of emissions controls, and conducting historical health assessments. Support from federal and other sources may be necessary to continue monitoring operation.

The research community and regulatory agencies should address the development and deployment of the most useful CO monitoring network for the purposes of planning and modeling. Monitors that will provide benefits in terms of assessing transportation impacts should be retained. In addition, to accurately assess local air pollution, permanent monitors need to reflect changes in growth and development patterns. Furthermore, as national interest in toxic air contaminant concentrations and environmental justice issues continues to increase, regulatory and research communities should evaluate placing additional monitoring stations to address those issues.

Twenty years down the road, most large cities might face the challenge of attaining O_3 and PM standards. Research conducted today addressing the tail end of the CO problem may provide important insights into the monitoring, modeling, and management of these other pollutants.

References

ADEC (Alaska Department of Environmental Conservation). 2001. State Air Quality Control Plan, Vol. 2. Analysis of Problems, Control Actions, Section III.C: Fairbanks Transportation Control Program. Adapted July 27, 2001. ADEC, Juneau, AK.

Air Improvement Resource, Inc. 1999. Office of the State Auditor Final Report: 1999 Audit of the Colorado AIR Program, November 22, 1999. Novi, MI: Air Improvement Resource, Inc.

Albu, S. 2002. Cold Start Emissions Reduction Strategies. Presentation at the Fifth Meeting on Carbon Monoxide Episodes in Meteorological and Topographical Problem Areas, April 8, 2002, Irvine, CA.

Akland, G.G., T.D. Hartwell, T.R. Johnson, and R.W. Whitmore. 1985. Measuring human exposure to carbon monoxide in Washington, D.C., and Denver, Colorado, during the winter of 1982-1983. Environ. Sci. Technol. 19(10):911-918.

Allred, E.N., E.R. Bleecker, B.R. Chaitman, T.E. Dahms, S.O. Gottlieb, J.D. Hackney, M. Pagano, R.H. Selvester, S.M. Walden, and J. Warren. 1989a. Short-term effects of carbon monoxide exposure on the exercise performance of subjects with coronary artery disease. N. Engl. J. Med. 321(21):1426-1432.

Allred, E.N., E.R. Bleecker, B.R. Chaitman, T.E. Dahms, S.O. Gottlieb, J.D. Hackney, D. Hayes, M. Pagano, R.H. Selvester, S.M. Walden, et al. 1989b. Acute effects of carbon monoxide exposure on individuals with coronary artery disease. Res. Rep. Health Eff. Inst. (25):1-79.

ARCADIS. 2003. Particulate Matter (PM2.5 and PM10) Apportionment for On-Road Mobile Sources. Final Report. Project HR-25-18. Prepared for National Cooperative Research Program, Transportation Research Board, National Research Council, Washington, DC, by ARCADIS G&M, Durham, NC, in association with Desert Research Institute, Reno, NV. January 2003.

Atkinson, R. 1994. Gas-Phase Tropospheric Chemistry of Organic Compounds. Journal of Physical and Chemical Reference Data, Monograph No. 2. Washington, DC: American Chemical Society.

Atkinson, R. 2000. Atmospheric chemistry of VOCs and NO_x. Atmos. Environ. 34(12):2063-2101.

Bae, C.C. 1993. Air quality and travel behavior: Untying the knot. J. Am. Plan. Assoc. 59(1): 65-74.

Beard, R.R., and G.A. Wertheim. 1967. Behavioral impairment associated with small doses of carbon monoxide. Am. J. Public Health Nations Health 57(11):2012-2022.

Beckett, W.S. 1994. The epidemiology of occupational asthma. Eur. Respir. J. 7(1):161-164.

Benignus, V.A., D.A. Otto, J.D. Prah, and G. Benignus. 1977. Lack of effects of carbon monoxide on human vigilance. Percept. Mot. Skills 45(3 Pt.1):1007-1014.

Benson, J.D., V. Burns, R.A. Gorse, A.M. Hochhauser, W.J. Koehl, L.J. Painter, and R.M. Reuter. 1991. Effects of Gasoline Sulfur Level on Mass Exhaust Emissions - Auto/Oil Air Quality Improvement Research Program. SAE 912323. Warrendale, PA: Society of Automotive Engineers.

Big Dig. 2000. The Big Dig. The Cental Artery/Tunnel Project, Boston, MA [Online]. Available: http://www.bigdig.com/ [accessed April 30, 2003].

Bowen, J.L., P.A. Walsh, and R.C. Henry. 1996. Analysis of Data From Lynwood Carbon Monoxide Study. Final Report A032-184. Prepared by Desert Research Institute, Reno, NV, for Research Division, California Air Resources Board, Sacramento, CA.

Bowling, S.A. 1984. Meteorological Factors responsible for high CO levels in Alaskan cities. Environmental Research Laboratory, U.S. Environmental Protection Agency, Corvallis, OR.

Bowling, S.A. 1986. Climatology of high-latitude air pollution as illustrated by Fairbanks and Anchorage, Alaska. J. Climate Appl. Met. 25:22-34.

Brunekreef, B. 1997. Air pollution from truck traffic and lung function in children living near motorways. Epidemiology 8(3):298-303.

Buckeridge, D.L., R. Glazier, B.J. Harvey, M. Escobar, C. Amrhein, and J. Frank. 2002. Effect of motor vehicle emissions on respiratory health in an urban area. Environ. Health Perspect. 110(3):293-300.

Burchell, R.W., G. Lowenstien, W.R. Dolphin, C.C. Galley, A. Downs, S. Seskin, K.G. Still, and T. Moore. 2002. Costs of Sprawl – 2000. Transit Cooperative Research Program Report 74. Washington, DC: National Academy Press.

Burnett, R.T., R.E. Dales, J.R. Brook, M.E. Raizenne, and D. Krewski. 1997. Association between ambient carbon monoxide levels and hospitalizations for congestive heart failure in the elderly in 10 Canadian cities. Epidemiology 8(2):162-167.

Calexico. 1999. Calexico East Port of Entry: 1999 Crosser's Report. Port of Entry, City of Calexico [Online]. Available: http://calexico.ca.gov/port_of_entry.htm [accessed Nov. 20, 2002].

Cambridge Systematics, Inc. 2001. Quantifying Air-Quality and Other Benefits and Costs of Transportation Control Measures. National Cooperative Highway Research Program Report 462. Washington, DC: National Academy Press.

Canale, R.P., S.R. Winegarden, C.R. Carlson, and D.L. Miles. 1978. General Motors Phase II Catalyst System. SAE Trans. 87:843-852.

CARB (California Environmental Protection Agency Air Resources Board). 1999. Air Quality Impacts of the Use of Ethanol in California Reformulated Gasoline. Appendix C. Baseline and Future Air Quality Concentrations. December 1999 [Online]. Available: http://www.arb.ca.gov/cbg/ethanol/ethfate/Airq/AppCf.pdf [accessed Nov. 20, 2002].

CARB (California Environmental Protection Agency Air Resources Board). 2002. Mobile Source Emissions Inventory Program. On-Road: EMFAC 2002 [Online]. Available: http://www.arb.ca.gov/msei/msei.htm [accessed Nov. 20, 2002].

CARB (California Environmental Protection Agency Air Resources Board). 2003a. California's Air Quality History—Key Events [Online]. Available: http://www.arb.ca.gov/html/ brochure/history.htm [accessed Feb. 3, 2003.]

CARB (California Environmental Protection Agency Air Resources Board). 2003b. Glossary of Air Pollution Terms [Online]. Available: http://www.arb.ca. gov/ html/ gloss.htm [accessed April 30, 2003].

Carter, W.P.L. 1998. Summary of Status of VOC Reactivity Estimates. Statewide Air Pollution Research Center and College of Engineering, Center for Environmental Research and Technology, University of California, Riverside, CA [Online]. Available: ftp://ftp.cert.ucr.edu/pub/carter/pubs/rcttab.pdf [accessed Nov. 20, 2002].

Chandra S., J.R. Ziemke, W. Min, and W.G. Read. 1998. Effects of 1997-1998 El Nino on tropospheric ozone and water vapor. Geophys. Res. Lett. 25(20): 3867-3870.

City of Fort Collins. 2001. City of Fort Collins Air Quality Action Plan Update (March 1999, Revised March 20, 2001). Air Quality Action Plan – 2000-2003. Air Quality Plans and Policy [Online]. Available: http://www.ci.fort-collins.co.us/airquality/plans-policies.php [accessed April 30, 2003].

Chrysler Corporation. 1998. Emission and Fuel Economy Regulation. Environmental & Energy Planning. CIMS 482-00-71. Chrysler Corporation, Auburn Hills, MI.

Cobb, N., and R.A. Etzel. 1991. Unintentional carbon monoxide-related deaths in the United States, 1979 through 1988. JAMA 266(5):659-663.

Coburn, R.F. 1970. Enhancement by phenobarbital and diphenylhydantoin of carbon monoxide production in normal man. N. Engl. J. Med. 283(10):512-515.

Colorado Department of Public Health and Environment. 2000. Carbon Monoxide Redesignation Request and Maintenance Plan for the Denver Metropolitan Area. Colorado Department of Public Health and Environment, Air Pollution Control Division, Denver, CO. [Online]. Available: http://www.cdphe.state.co.us/ap/down/sipdenco.pdf [accessed Nov. 20, 2002].

Dana, G. 2002. NAS Committee on Carbon Monoxide Episodes in Meteorological and Topographical Problem Areas. Presentation at the Sixth meeting on Carbon Monoxide Episodes in Meteorological and Topographical Problem Areas, July 16, 2002, Missoula, MT.

Davis, S.C. 1997. Transportation Energy Data Book: Edition 17. ORNL-6919. Center for Transportation Analysis, Oak Ridge National Laboratory, Oak Ridge, TN.

Davis, S.C. 2001. Transportation Energy Data Book: Edition 21. ORNL-6966. Center for Transportation Analysis, Oak Ridge National Laboratory, Oak Ridge, TN.

DHEW (U.S. Department of Health, Education, and Welfare). 1970. Air Quality Criteria for Carbon Monoxide. National Air Pollution Control Administration Pub. No. AP-42. Washington, DC: U.S. Govt. Print. Off. March 1970.

DOE/DOD (U.S. Department of Energy and U.S. Department of Defense). 2002. Joint Urban 2003, Atmospheric Dispersion Study, Oklahoma City, July 2003. Chemical and Biological National Security Program, National Nuclear Security Administration, U.S. Department of Energy, and Defense Threat Reduction Agency, U.S. Department of Defense.

DOE/EPA (U.S. Department of Energy and U.S. Environmental Protection Agency). 2002. Hybrid Vehicles [Online]. Available: http://www.fuel economy.gov/feg/hybrid_sbs.shtml [accessed Nov. 19, 2002].

DOT (U.S. Department of Transportation). 2000. Transportation Conformity: A Basic Guide for State and Local Officials. Federal Highway Administration, U.S. Department of Transportation. Revised June 19, 2000 [Online]. Available: http://www.fhwa.dot.gov/environment/conformity/basic_gd.htm [accessed Nov. 21, 2002].

DOT/EPA (U.S. Department of Transportation and U.S. Environmental Protection Agency). 2002. Transportation and Air Quality Public Information Initiative, It All Adds Up to Cleaner Air, Research Findings. U.S. Department of Transportation and U.S. Environmental Protection Agency [Online]. Available: http://www.epa.gov/oms/transp/sti/res-1pgr.pdf [accessed Nov. 21, 2002].

DOT/EPA (U.S. Department of Transportation and U.S. Environmental Protection Agency). 2003. It All Adds Up to Cleaner Air. U.S. Department of Transportation and U.S. Environmental Protection Agency [Online]. Available: http://www. italladdsup. gov/ [accessed April 29, 2003].

Dulla, R., and P. Heirigs. 2002. Problems With Forecasted CO Emission Rates in MOBILE6. Memo to Wayne Elson, Region 10, U.S. Environmental Protection Agency, from R. Dulla and P. Heirings, Sierra Research, Sacramento, CA. September 4, 2002.

Eastman, J.L., R.A. Pielke, and W.A. Lyons. 1995. Comparison of lake-breeze model simulations with tracer data. J. Appl. Meteorol. 34(6):1398-1418.

Eisinger, D.S., K. Dougherty, D.P.Y. Chang, T. Kear, and P.F. Morgan. 2002. A reevaluation of carbon monoxide: Past trends, future concentrations, and implications for conformity "hot-spot" policies. J. Air Waste Manage. Assoc. 52(9):1012-1025.

EIA (U.S. Energy Information Agency). 2001. Annual Energy Outlook 2002 With Projections to 2020. DOE/EIA-0383(2003). U.S. Department of Energy, Washington, DC. December 2001 [Online]. Available: http://tonto.eia.doe.gov/FTPROOT/forecasting/forecasting.htm [accessed April 29, 2003].

EIA (U.S. Energy Information Agency). 2003. Annual Energy Outlook 2003 With Projections to 2025. DOE/EIA-0383(2003). U.S. Department of Energy, Washington, DC [Online]. Available: http://www.eia.doe.gov/oiaf/aeo.html [accessed April 29, 2003].

ENVIRON. 1998. Performance Audit of the Colorado AIR Program, March 1998. Final Report. Prepared for Office of State Auditor, State of Colorado, Denver, CO, by ENVIRON International Corporation, Novato, CA [Online]. Available: http://www.state.co.us/gov_dir/audit_dir/1998/98perf/perf98.htm [accessed Nov. 22, 2002].

EPA (U.S. Environmental Protection Agency). 1973. The National Air Monitoring Program: Air Quality and Emissions Trends. Annual Report. EPA-450/1-73-001a-b. Office of Air Quality Planning and Standards, U.S. Environmental Protection Agency, Research Triangle Park, NC.

EPA (U.S. Environmental Protection Agency). 1976. Monitoring and Air Quality Trends Report, 1974. EPA-450/1-76-001. Office of Air Quality Planning and Standards, Monitoring and Data Analysis Division, U.S. Environmental Protection Agency, Research Triangle Park, NC.

EPA (U.S. Environmental Protection Agency). 1979. Air Quality Criteria for Carbon Monoxide. EPA-600/8-79-022. Environmental Criteria and Assessment Office, Office of Research and Development, U.S. Environmental Protection Agency, Research Triangle Park, NC.

EPA (U.S. Environmental Protection Agency). 1992. Guideline for Modeling Carbon Monoxide from Roadway Intersections. EPA 454/R-92-005. NTIS PB 93-210391. Office of Air Quality Planning and Standards, U.S. Environmental Protection Agency, Research Triangle Park, NC.

EPA (U.S. Environmental Protection Agency). 1995. Compilation of Air Pollutant Emission Factors, AP-42, Fifth Ed. Office of Air and Radiation, U.S. Environmental Protection Agency [Online]. Available: http://www.epa.gov/ttn/chief/ap42/index.html [accessed Nov. 21, 2002].

EPA (U.S. Environmental Protection Agency). 1996. Regulatory Impact Analysis: Federal Test Procedure Revisions. Office of Air and Radiation, Office of Mobile Sources, U.S. Environmental Protection Agency, Ann Arbor, MI. August 15, 1996.

EPA (U.S. Environmental Protection Agency). 1997a. Final Report to Congress on Benefits and Costs of the Clean Air Act, 1970 to 1990. EPA 410-R-97-002. Office of Air and Radiation, U.S. Environmental Protection Agency. Cincinnati, OH: National Service Center for Environmental Publications. October 1997.

EPA (U.S. Environmental Protection Agency). 1997b. Survey of Episodic Control Programs. Office of Transportation and Air Quality, U.S. Environmental Protection Agency [Online]. Available: http://www.epa.gov/OMSWWW/reports/episodic/study.htm [accessed Nov. 21, 2002].

EPA (U.S. Environmental Protection Agency). 1997c. Episodic Control Programs. Environmental Fact Sheet. EPA420-F-97-022. Office of Mobile Sources, Office of Air and Radiation, U.S. Environmental Protection Agency. December 1997 [Online]. Available: http://www.epa.gov/otaq/transp/42097022.pdf [accessed Nov. 21, 2002].

EPA (U.S. Environmental Protection Agency). 1997d. Guidance on Incorporating Voluntary Mobile Source Emission Reduction Programs in State Implementation Plans (SIPs). Fact Sheet. October 24, 1997. Office of Transportation and Air Quality, U.S. Environmental Protection Agency [Online]. Available: http://www.epa.gov/oms/transp/trancont/vmep-fs.txt [accessed Nov. 21, 2002].

EPA (U.S. Environmental Protection Agency). 1998a. Final Guidance for Incorporating Environmental Justice Concerns in EPA's NEPA Compliance Analyses. Office of Federal Activities, U.S. Environmental Protection Agency. April 1998 [Online]. Available: http://www.epa.gov/Compliance/resources/policies/ej/ej_guidance_nepa_epa0498.pdf [accessed Nov. 20, 2002].

EPA (U.S. Environmental Protection Agency). 1998b. Traffic Flow Improvements. Transportation Control Measure. Transportation and Air Quality TCM Technical Overviews. EPA420-S-98-012. Transportation and Air Quality Planning, U.S. Environmental Protection Agency, Washington, DC. July 1998 [Online]. Available: http://www.epa.gov/orcdizux/transp/publicat/pub_tech.htm [accessed Nov. 20, 2002].

EPA (U.S. Environmental Protection Agency). 1999. Final Report to Congress on Benefits and Costs of the Clean Air Act, 1990 to 2010. EPA 410-R-99-001. Office of Air and Radiation, U.S. Environmental Protection Agency. November 1999.

EPA (U.S. Environmental Protection Agency). 2000a. Air Quality Criteria for Carbon Monoxide. EPA/600/P-99/001F. Office of Research and Development, National Center for Environmental Assessment, Washington, DC. June 2000 [Online]. Available: http://www.epa.gov/NCEA/pdfs/coaqcd.pdf [accessed Nov. 20, 2002].

EPA (U.S. Environmental Protection Agency). 2000b. National Air Pollutant Emission Trends, 1900-1998. EPA-454/R-00-002. Office of Air Quality Planning and Standards, U.S. Environmental Protection Agency, Research

Triangle Park, NC [Online]. Available: http://www.epa.gov/ttn/chief/trends/trends98/ [accessed Nov. 20, 2002].

EPA (U.S. Environmental Protection Agency). 2000c. Draft Technical Support Document: Control of Emissions of Hazardous Air Pollutants from Motor Vehicles and Motor Vehicle Fuels. EPA-420/D-00-003. Assessment and Standards Division, Office of Transportation and Air Quality, U.S. Environmental Protection Agency [Online]. Available: http://www.epa.gov/otaq/regs/toxics/ d00003.pdf [accessed Nov. 20, 2002].

EPA (U.S. Environmental Protection Agency). 2001a. National Air Quality and Emissions Trends Report, 1999. EPA/454/R-01-004. Office of Air Quality Planning and Standards, Emissions Monitoring and Analysis Division, Air Quality Trends Analysis Group, U.S. Environmental Protection Agency, Research Triangle Park, NC. March 2001 [Online]. Available: http://www.epa.gov/oar/ aqtrnd99/ [accessed Nov. 19, 2002].

EPA (U.S. Environmental Protection Agency). 2001b. Latest Findings on National Air Quality: 2000 Status and Trends. EPA-454/K-01-002. Office of Air Quality Planning and Standards, U.S. Environmental Protection Agency, Research Triangle Park, NC [Online]. Available: http://www.epa.gov/oar/aqtrnd00/brochure/ 00brochure.pdf [accessed April 30, 2003].

EPA (U.S. Environmental Protection Agency). 2001c. The Projection of Mobile Source Air Toxics from 1996 to 2007: Emissions and Concentrations. Draft Report. EPA-420/R-01-038. Office of Air Quality Planning and Standards, U.S. Environmental Protection Agency, Research Triangle Park, NC [Online]. Available: http://www.epa.gov/orcdizux/regs/toxics/r01038.pdf [accessed Nov. 20, 2002].

EPA (U.S. Environmental Protection Agency). 2001d. EPA's Commitment to Environmental Justice. Memorandum to Assistant Administrators, General Counsel, Inspector General, Chief Financial Officer, Associate Administrators, Regional Administrators, Office Directors, from C. Todd Whitman, Administrator, U.S. Environmental Protection Agency. August 9, 2001 [Online]. Available: http://www.epa.gov/Compliance/resources/policies/ej/index.html [accessed Nov. 20, 2002].

EPA (U.S. Environmental Protection Agency). 2001e. TCM Programs Organized by Program Category. Transportation Control Measures Program Information Directory, U.S. Environmental Protection Agency [Online]. Available: http://yosemite.epa.gov/aa/tcmsitei.nsf/9bd6f3b7217f80c28525652f0053e105/fed7fcb81187584585256 5800066038f?OpenDocument [accessed Nov. 22, 2002].

EPA (U.S. Environmental Protection Agency). 2001f. Our Built and Natural Environments: A Technical Review of the Interactions between Land Use, Transportation, and Environmental Quality. EPA 231-R-01-002. Development, Community, and Environmental Division, U.S. Environmental Protection

Agency. January 2001 [Online]. Available: http://www.epa.gov/piedpage/pdf/ built.pdf [accessed April 29, 2003].

EPA (U.S. Environmental Protection Agency). 2001g. EPA Guidance: Improving Air Quality Through Land Use Activities. EPA420-R-01-001. Transportation and Regional Programs Division, Office of Transportation and Air Quality, U.S. Environmental Protection Agency. January 2001 [Online]. Available: http://www.epa.gov/otaq/transp/trancont/r01001.pdf [accessed April 30, 2003].

EPA (U.S. Environmental Protection Agency). 2002a. National Air Quality Status and Trends 2001. Office of Air and Radiation, U.S. Environmental Protection Agency [Online]. Available: http://www.epa.gov/oar/aqtrnd01/ [accessed March 22, 2003].

EPA (U.S. Environmental Protection Agency). 2002b. Air Trends Reports. Air Quality Planning and Standards, Office of Air and Radiation, U.S. Environmental Protection Agency [Online]. Available: http://www.epa.gov/airtrends/reports.html [accessed April 30, 2003].

EPA (U.S. Environmental Protection Agency). 2002c. Health Assessment Document for Diesel Engine Exhaust. EPA/600/8-90/057F. National Center for Environmental Assessment, Office of Research and Development, U.S. Environmental Protection Agency, Washington, DC.

EPA (U.S. Environmental Protection Agency). 2002d. Whitman Announces U.S.-Mexico Border Air Quality Strategy. EPA Newsroom, November 26, 2002 [Online]. Available: http://www.epa.gov/epahome/headline_112702.htm [accessed April 30, 2003].

EPA (U.S. Environmental Protection Agency). 2002e. Reports from the Work Groups of the Mobile Source Technical Review Subcommittee for the MSTRS Meeting of February 13, 2002. Office of Air and Radiation, U.S. Environmental Protection Agency [Online]. Available: http://www.epa.gov/air/caaac/wgrpts202.pdf [accessed Nov. 21, 2002].

EPA (U.S. Environmental Protection Agency). 2002f. Action Days. Air Now. Office of Air Quality Planning and Standards, Office of Air and Radiation, U.S. Environmental Protection Agency [Online]. Available: http://www.epa.gov/airnow/action.html [accessed Nov. 21, 2002].

EPA (U.S. Environmental Protection Agency). 2002g. National Ambient Air Monitoring Strategy: Summary Document. Final Draft for Comment. Ambient Monitoring Technology Information Center, Office of Air Quality Planning and Standards, Office of Air and Radiation, U.S. Environmental Protection Agency. September 1, 2002 [Online]. Available: http://www.epa.gov/ttn/amtic/files/ambient/monitorstrat/summary.pdf [accessed Nov. 21, 2002].

EPA (U.S. Environmental Protection Agency). 2002h. AirData: Access to Air Pollution Data. Office of Air and Radiation, U.S. Environmental Protection Agency [Online]. Available: http://www.epa.gov/air/data/ [accessed Nov. 21, 2002].

EPA (U.S. Environmental Protection Agency). 2002i. User Guide to MOBILE6.1 and MOBILE6.2: Mobile Source Emissions Factor Model. EPA-420-R-02-028. Assessment and Standards Division, Office of Transportation and Air Quality, Office of Air and Radiation, U.S. Environmental Protection Agency [Online]. Available: http://www.epa.gov/otaq/m6.htm [accessed Nov. 21, 2002].

EPA (U.S. Environmental Protection Agency). 2002j. MOBILE6 Vehicle Emission Modeling Software. Office of Transportation and Air Quality Modeling, U.S. Environmental Protection Agency [Online]. Available: http://www.epa.gov/otaq/m6.htm [accessed Nov. 21, 2002].

EPA (U.S. Environmental Protection Agency). 2002k. NONROAD Model (Off-Road Vehicles, Equipment, and Vehicles). Office of Transportation and Air Quality Modeling, U.S. Environmental Protection Agency [Online]. Available: http://www.epa.gov/otaq/nonrdmdl.htm [accessed Nov. 21, 2002].

EPA (U.S. Environmental Protection Agency). 2002l. The Plain English Guide To The Clean Air Act. Office of Air Quality Planning and Standards, Office of Air and Radiation, U.S. Environmental Protection Agency [Online]. Available: http://www.epa.gov/oar/oaqps/peg_caa/pegcaa10.html [accessed April 30, 2003].

EPA (U.S. Environmental Protection Agency). 2003a. The National-Scale Air Toxics Assessment. Technology Transfer Network. National Air Toxics Assessment. Office of Air and Radiation. U.S. Environmental Protection Agency [Online]. Available: http://www.epa.gov/ttn/atw/nata/ [accessed March 23, 2003].

EPA (U.S. Environmental Protection Agency). 2003b. I/M Best Practices - Key Terms. Cars and Light Trucks. Inspection & Maintenance (I/M). Office of Transportation and Air Quality, U.S. Environmental Protection Agency [Online]. Available: http://epa.gov/otaq/epg/keyterm.htm [accessed April 30, 2003].

EPA (U.S. Environmental Protection Agency). 2003c. Federal and California Exhaust and Evaporative Emission Standards for Light-Duty Vehicles and Light-Duty Trucks. Cars and Light Trucks. Office of Transportation and Air Quality, U.S. Environmental Protection Agency [Online]. Available: http://www.epa.gov/otaq/stds-ld.htm [accessed April 30, 2003].

FAA (Federal Aviation Administration). 2002. Emissions and Dispersion Modeling System. EDMS 4.01 Software Update. Federal Aviation Administration, Office of Environment and Energy [Online]. Available: http://www.aee.faa.gov/edms/edms401.htm [accessed Nov. 21, 2002].

FHWA (Federal Highway Administration). 1995. Transportation Control Measure Analysis, Transportation Control Measures Analyzed for the Washington Region's 15 Percent Rate of Progress Plan. Metropolitan Planning Technical Report No. 5. Federal Highway Administration, U.S. Department of Transportation, Washington, DC. February 1995.

FHWA (Federal Highway Administration). 1998. FHWA Actions to Address Environmental Justice in Minority Populations and Low-Income Populations. Order 6620.23. Federal Highway Administration, U.S. Department of Transportation, Washington, DC. December 2, 2002 [Online]. Available: http://www.fhwa.dot.gov/legsregs/directives/orders/6640_23.htm [accessed Nov. 20, 2002].

Flachsbart, P.G. 1999. Human exposure to carbon monoxide from mobile sources. Chemosphere - Global Change Science 1(1-3):301-329.

Flachsbart, P.G., G.A. Mack, J.E. Howes, and C.E. Rodes. 1987. Carbon monoxide exposures of Washington commuters. JAPCA 37(2):135-142.

Flaum, J.B., S.T. Rao, and I.G. Zurbenko. 1996. Moderating the influence of meteorological conditions on ambient ozone concentrations. J. Air Waste Manage. 46(1):35-46.

Frey, H.C., and J. Zheng. 2002. Probabilistic analysis of driving cycle-based highway emission factors. Environ. Sci. Technol. 36(23):5184-5191.

Fujita, E.M., J.G. Watson, J.C. Chow, N.F. Robinson, L. Richards, and N. Kumar. 1998. Northern Front Range Air Quality Study, Volume C: Source Apportionment and Simulation Methods and Evaluation. Final Report. Prepared by Desert Research Institute, Reno, NV, for Colorado State University, Fort Collins, CO. June 30, 1998.

Fullerton, N., ed. 2002. FSL in Review: Fiscal Year 2001 Accomplishments, Projections for 2002. Boulder, CO: Forest Systems Laboratory, Environmental Research Laboratories, National Oceanic and Atmospheric Administration, U.S. Department of Commerce.

Fulton, W., R. Pendall, M. Nguyen, and A. Harrison. 2001. Who Sprawls Most? How Growth Patterns Differ Across the U.S. Center on Urban and Metropolitan Policy, The Brookings Institution, Washington, DC. July 2001 [Online]. Available: http://www.brook.edu/dybdocroot/es/urban/fulton-pendall.htm [accessed Nov. 22, 2002].

Geer, I.W., ed. 1996. Glossary of Weather and Climate: With Related Oceanic and Hydrologic Terms. Boston, MA: American Meteorological Society. 272 pp.

Gibbons, R. 2002. Environmental Regulatory Statistics. Presentation at the Sixth Meeting on Carbon Monoxide Episodes in Meteorological and Topographical Problem Areas, July 17, 2002, Missoula, MT.

Granell, J. 2002. Model Year Distribution and Vehicle Technology Composition of the Onroad Fleet as a Function of Vehicle Registration Data and Site Location Characteristics. Ph.D. Dissertation. Georgia Institute of Technology, Atlanta, GA. November 2002 [Online]. Available: http://transaq.ce.gatech.edu/guensler/publications/theses/theses.html [accessed May 21, 2003].

Guay, G. 2001. Fairbanks CO Saturation Study. Presentation at the Second Meeting on Carbon Monoxide Episodes in Meteorological and Topographical Problem Areas, August 8, 2001, Fairbanks, AK.

Guensler, R. 1989. Preparation of Alternative Emission Control Plans for Surface Coating Operations. California Air Resources Board, Sacramento, CA. Paper No. 89-118B.3. In Proceedings: 82nd Annual Meeting and Exhibition of the Air & Waste Management Association, June 25-30, 1989, Anaheim, CA, Vol. 7 [Online]. Available: http://transaq.ce.gatech.edu/guensler/publications/proceedings/proceedings.html [accessed Nov. 22, 2002].

Guensler, R. 1998. Increasing vehicle occupancy in the United States. Pp. 127-155 in L'Avenir Des Déplacements en Ville (The Future of Urban Travel), O. Andan et al., eds. Lyon, France: Laboratoire d'economie des transports [Online]. Available: http://transaq.ce.gatech.edu/guensler/publications/proceedings/proceedings.html [accessed Nov. 22, 2002].

Guensler, R. 2000. TRANS/AQ 2000. Transportation and Air Quality. Courseware CD-ROM; Version 1.0. Georgia Institute of Technology, Atlanta, GA [Online]. Available: http://transaq.ce.gatech.edu/ [accessed Nov. 22, 2002].

Guensler, R., S. Shaheen, F. Mar, and C. Yee. 1998. Conformity Policy: Air Quality Impact Assessment for Local Transportation Projects. Paper UCD-ITS-RR-98-02. Institute of Transportation Studies, University of California, Davis, CA [Online]. Available: http://transaq.ce.gatech.edu/guensler/publications/reports/reports.html [accessed March 20, 2003]

Hallock-Waters, K.A., B.G. Doddridge, R.R. Dickerson, S. Spitzer, and J.D. Ray. 1999. Carbon monoxide in the U.S. Mid-Atlantic troposphere: Evidence for a decreasing trend. Geophys. Res. Lett. 26(18):2861-2864.

Hanna, S.R. 1978. Accuracy of dispersion models: A position paper of the AMS 1977 committee on atmospheric turbulence and diffusion. Bull. Am. Meteorol. Soc. 59(8):1025-1026.

Hargesheimer, N. 2001. Fairbanks: The Community and CO. Presentation at the First Meeting on Carbon Monoxide Episodes in Meteorological and Topographical Problem Areas, June 4, 2001, Washington, DC.

Harvey, G., and E. Deakin. 1993. A Manual of Regional Transportation Modeling Practice for Air Quality Analysis, Version 1.0. National Association of Regional Councils.

Hoek, G., B. Brunekreef, S. Goldbohm, P. Fisher, and P.A. van den Brandt. 2002. Association between mortality and indicators of traffic-related air pollution in the Netherlands: A cohort study. Lancet 360(9341):1203-1209.

Holmes, K.J., and A.G. Russell. 2001. Improving mobile-source emissions modeling. EM (February):20-28.

Horvath, S.M., T.E. Dahms, and J.F. O'Hanlon. 1971. Carbon monoxide and human vigilance: A deleterious effect of present urban concentrations. Arch. Environ. Health 23(5):343-347.

Howitt, A.M., and E.M. Moore. 1999. Implementing the transportation conformity regulations. TR News 202(May-June):15-23, 41.

IMRC (California Inspection and Maintenance Review Committee). 2000. Smog

Check II Evaluation. California Inspection and Maintenance Review Committee, Sacramento, CA.

Joy, R.W., T.P. Kear, and M.P. Valdez. 1995. Analysis of the Probablity of Continued Carbon Monoxide Attainment in Puget Sound. SR95-05-04. Prepared for Puget Sound Air Pollution Control Agency, by Sierra Research, Inc., Sacramento, CA.

King, C.W. 1991. Selecting Meteorological Variables as Input for Statistical Carbon Monoxide Forecast Models. Paper 91-54.2. Presentation at the 84th Annual Meeting, Air and Waste Management Association, June 16-21, 1991, Vancouver, BC.

King, R. 2001. Recent Efforts by the State of Alaska to Help Fairbanks to Attain the NAAQS. Presentation at the First Meeting on Carbon Monoxide Episodes in Meteorological and Topographical Problem Areas, June 4, 2001, Washington, D.C.

Kirchstetter, T.W., B.C. Singer, R.A. Harley, G.R. Kendall, and M. Traverse. 1999. Impact of California reformulated gasoline on motor vehicle emissions. 1. Mass emissions rates. Environ. Sci. Technol. 33(2):318-328.

Kuebler, J., H. van den Bergh, and A.G. Russell. 2001. Long-term trends of primary and secondary pollutant concentrations in Switzerland and their response to emission controls and economic changes. Atmos. Environ. 35(8):1351-1363.

Lawson, D.L. 2002. Mobile Source Carbon Monoxide Emissions. Presentation at the Fifth Meeting on Carbon Monoxide Episodes in Meteorological and Topographical Problem Areas, April 8, 2002, Irvine, CA.

Lawson, D.R., P.J. Groblicki, D.H. Stedman, G.A. Bishop, and P.L. Guenther. 1990. Emissions from in-use motor vehicles in Los Angeles: A pilot study of remote sensing and inspections and maintenance program. J. Air Waste Manage. Assoc. 40(8):1095-1105.

Lena, T.S., V. Ochieng, M. Carter, J. Holguin-Veras, and P.L. Kinney. 2002. Elemental carbon and PM2.5 levels in an urban community heavily impacted by truck traffic. Environ. Health Perspect. 110(10):1009-1015.

Lyons, W.A., C.J. Tremback, and R.A. Pielke. 1995. Applications of the Regional Atmospheric Modeling System (RAMS) to provide input to photochemical grid models for the Lake Michigan Ozone Study (LMOS). J. Appl. Meteorol. 34(8):1762-1786.

Mahrer, Y., and R.A. Pielke. 1977. The effects of topography on sea and land breezes in a two-dimensional numerical model. Mon. Wea. Rev. 105:1151-1162.

Marr, L.C., G.C. Morrison, W.W. Nazaoff, and R.A. Harley. 1998. Reducing the risk of accidental death due to vehicle-related carbon monoxide poisoning. J. Air Waste Manage. Assoc. 48(10):899-906.

McFarland, R.A. 1970. The effects of exposure to small quantities of carbon monoxide on vision. Ann. N.Y. Acad. Sci. 174(1):301-312.

McFarland, R.A. 1973. Low level exposure to carbon monoxide and driving performance. Arch. Environ. Health 27(6):355-359.

Moolgavkar, S.H., E.G. Luebeck, and E.L. Anderson. 1997. Air pollution and hospital admissions for respiratory causes in Minneapolis-St. Paul and Birmingham. Epidemiology 8(4):364-370.

Morris, S. 2001. Carbon Monoxide in Anchorage – What We've Learned in Twenty Years. Presentation at the Second Meeting on Carbon Monoxide Episodes in Meteorological and Topographical Problem Areas, August 8, 2001, Fairbanks, AK.

Morris, R.D., E.N. Naumova, and R.L. Munasinghe. 1995. Ambient air pollution and hospitalization for congestive heart failure among elderly people in seven large U.S. cities. Am. J. Public Health 85(10):1361-1365.

Morris, S.S., and L. Taylor, Jr. 1998. Winter 1997-98 Anchorage Carbon Monoxide Saturation Monitoring Study. Air Quality Program Report. Municipality of Anchorage, Dept. of Health and Human Services. September 1998.

Mott, J.A., M.I. Wolfe, C.J. Alverson, S.C. MacDonald, C.R. Bailey, L.B. Ball, J.E. Moorman, J.H. Somers, D.M. Mannino, and S.C. Redd. 2002. National vehicle emissions policies and practices and declining US carbon monoxide-related mortality. J. Am. Med. Assoc. 288(8):988-995.

Mulawa, P.A., S.H. Cadle, K. Knapp, R. Zweidinger, R. Snow, R. Lucas, and J. Goldbach. 1997. Effect of ambient temperature and E-10 fuel on primary exhaust particulate matter emissions from light-duty vehicles. Environ. Sci. Technol. 31(5):1302-1307.

NARSTO (North American Research Strategy for Tropospheric Ozone). 2003. Particulate Matter Science for Policy Makers. A NARSTO Assessment. Final Report. EPRI 1007735. NARSTO Management Coordination Office, Pasco, WA. February 2003.

Neff, W. 2001. Meteorology of Air Quality in Colorado. Presentation at the Third Meeting on Carbon Monoxide Episodes in Meteorological and Topographical Problem Areas, November 9, 2001, Denver, CO.

Neff, W.D., and C.W. King. 1991. The Micrometeorology of Denver Carbon-Monoxide Episodes. Presentation at the 7th Joint Conference on Applications of Air Pollution Meteorology with AWMA, January 14-18, 1991, New Orleans, LA.

New York State Department of Environmental Conservation. 1999. New York State Implementation Plan. Carbon Monoxide Redesignation Request and Maintenance Plan for the New York Metropolitan Area. Division of Air Resources, Department of Environmental Conservation. August 1999.

Nininger, R.C. 1991. Determination of Source Contributions to High Ambient Carbon Monoxide Concentrations and Categorization of Carbon Monoxide Potential. Final Report. Contract No. A832-135. Prepared by AeroVironment, Inc., Monrovia, CA, for Research Division, California Air Resources Board, Sacramento, CA.

Norton, T., S. Tucker, R.E. Smith, and D.R. Lawson. 1998. The Northern Front Range Air Quality Study. EM (January):13-19.

NRC (National Research Council). 1977. Carbon Monoxide. Medical and Biologic Effects of Environmental Pollutants. Washington, DC: National Academy of Sciences.

NRC (National Research Council). 1999. Ozone-Forming Potential of Reformulated Gasoline. Washington, DC: National Academy Press.

NRC (National Research Council). 2000. Modeling Mobile-Source Emissions. Washington, DC: National Academy Press.

NRC (National Research Council). 2001. Evaluating Vehicle Emissions Inspection and Maintenance Programs. Washington, DC: National Academy Press.

NRC (National Research Council). 2002. The Ongoing Challenge of Managing Carbon Monoxide Pollution in Fairbanks, Alaska. Interim Report. Washington, DC: National Academy Press.

NSTC (National Science and Technology Council). 1997. Interagency Assessment of Oxygenated Fuels. National Science and Technology Council, Committee on Environment and Natural Resources, Office of Science and Technology Policy, Executive Office of the President of the United States.

NWS (National Weather Service). 2002. Environmental Modeling Center. National Weather Service, National Oceanic and Atmospheric Administration, U.S. Dept. of Commerce [Online]. Available: http://www.emc.ncep.noaa.gov/ [accessed April 30, 2003].

Pappas, G.P., R.J. Herbert, W. Henderson, J. Koening, B. Stover, and S. Barnhart. 2000. The respiratory effects of volatile organic compounds. Int. J. Occup. Environ. Health 6(1):1-8.

Parrish, D.D., M. Trainer, D. Hereid, E.J. Williams, K.J. Olszyna, R.A. Harley, J.F. Meagher, and F.C. Fehsenfeld. 2002. Decadal change in carbon monoxide to nitrogen oxide ratio in U.S. vehicular emissions. J. Geophys. Res. 107(12): ACH5.

Pielke, R.A. 2002. Mesoscale Meteorological Modeling, 2nd Ed. San Diego: Academic Press. 676 pp.

Pielke, R.A., and M. Uliasz. 1993. Influence of landscape variability on atmospheric dispersion. J. Air Waste Manage. 43(7):989-994.

Pielke, R.A., R.A. Stocker, R.W. Arritt, and R.T. McNider. 1991. A procedure to estimate worst-case air quality in complex terrain. Environ. Int. 17(6):559-574.

Pokharel, S.S., G.A. Bishop, and D.H. Stedman. 2002. An on-road motor vehicle emissions inventory for Denver: An efficient alternative to modeling. Atmos. Environ. 36(33):5177-5184.

Poloniecki, J.D., R.W. Atkinson, A.P. de Leon, and H.R. Anderson. 1997. Daily time series for cardiovascular hospital admissions and previous day's air pollution in London, UK. Occup. Environ. Med. 54(8):535-540.

Porter, P.S., S.T. Rao, I.G. Zurbenko, A.M. Dunker, and G.T. Wolff. 2001. Ozone air quality over North America: Part II - An analysis of trend detection and attribution techniques. J. Air Waste Manage. 51(2):283-306.

Prescott, G.J., G.R. Cohen, R.A. Elton, F.G. Fowkes, and R.M. Agius. 1998. Urban air pollution and cardiopulmonary ill health: A 14.5 year time series study. Occup. Environ. Med. 55(10):697-704.

Prinn, R.G., J. Huang, R.F. Weiss, D.M. Cunnold, P.J. Fraser, P.G. Simmonds, A. McCulloch, C. Harth, P. Salameh, S. O'Doherty, R.H.J. Wang, L. Porter, and B.R. Miller. 2001. Evidence for substantial variations of atmospheric hydroxyl radicals in the past two decades. Science 292(5523):1882-1887.

Ragazzi, R., and K. Nelson. 1999. The Impact of a 10% Ethanol Blended Fuel on the Exhaust Emissions of Tier 0 and Tier 1 Light Duty Gasoline Vehicles at 35° F. Colorado Department of Public Health and Environment, Denver, CO. March 26, 1999 [Online]. Available: http://www.cdphe.state.co.us/ap/down/oxyfuelstudy.PDF [accessed Nov. 22, 2002].

Rajan, S. 1993. Socio-Spatial Patterns of Vehicle Ownership in Southern California: Preliminary Study. University of California, Los Angeles, CA.

Ransel, D. 2002. The Carbon Monoxide Problem in Las Vegas, Nevada. Presentation at the Fifth Meeting on Carbon Monoxide Episodes in Meteorological and Topographical Problem Areas, April 8, 2002, Irvine, CA.

Raub, J.A., M. Mathieu-Nolf, N.B. Hampson, and S.R. Thom. 2000. Carbon monoxide poisoning – a public health perspective. Toxicology 145(1):1-14.

Reddy, P.J. 2000. Appendix M. Analysis of the Probability of the Carbon Monoxide Exceedance in Denver During the First Week of February for the Years 2002 through 2013 and Two Possible Levels of Oxygenates in Automotive Fuels Based on Historical Carbon Monoxide for 1975 through 1994. Pp. 575-580 in Technical Support Document: Carbon Monoxide Redesignation Request and Maintenance Plan for the Denver Metropolitan Area. Colorado Department of Public Health and Environment, Air Pollution Control Division, Denver, CO. January 4, 2000. [Online]. Available: http://apcd.state.co.us/documents/techdocs.html. [accessed March 26, 2003].

Reddy, P.J. 2001. Forecasting Carbon Monoxide and Air Quality Along Colorado's Front Range Urban Corridor. Presentation at the Third Meeting on Carbon Monoxide Episodes in Meteorological and Topographical Problem Areas, November 9, 2001, Denver, CO.

Regional Air Quality Center. 2002. Denver's Winter High Pollution Season. Regional Air Quality Center, Denver, CO [Online]. Available: http://www.raqc.org/winter/winter-act.htm [accessed Nov. 22, 2002].

Ritz, B. 2002. Ambient Air Pollution and Risk of Birth Defects in Southern California. Presentation at the Fifth Meeting on Carbon Monoxide Episodes in Meteorological and Topographical Problem Areas, April 8, 2002, Irvine, CA.

Ritz, B., and F. Yu. 1999. The effect of ambient carbon monoxide on low birth weight among children born in southern California between 1989 and 1993. Environ. Health Perspect. 107(1):17-25.

Ritz, B., F. Yu, G. Chapa, and S. Fruin. 2000. Effects of air pollution on preterm birth among children born in southern California between 1989 and 1993. Epidemiology 11(5):502-511.

Ritz, B., F. Yu, S. Fruin, G. Chapa, G.M. Shaw, and J.A. Harris. 2002. Ambient air pollution and risk of birth defects in southern California. Am. J. Epidemiol. 155(1):17-25.

Rodes, C., L. Sheldon, D. Whitaker, A. Clayton, K. Fitzgerald, J. Flanagan, F. DiGenova, S. Hering, and C. Frazier. 1998. Measuring Concentrations of Selected Air Pollutants Inside California Vehicles. Final Report. Prepared for California Environmental Protection Agency Air Resources Board, Sacramento, CA, and South Coast Air Quality Management District, Diamond Bar, CA. December 1998 [Online]. Available: http://www.arb.ca.gov/research/indoor/in-vehsm.htm [Nov. 20, 2002].

Sawyer, R.F., R.A. Harley, S.H. Cadle, J.M. Norbeck, R. Slott, and H.A. Bravo. 2000. Mobile sources critical review: 1998 NARSTO assessment. Atmos. Environ. 34(12-14):2161-2181.

SCAQMD (South Coast Air Quality Management District). 2000a. Air Quality Standards Compliance Report. Vol. 13, No. 12. South Coast Air Quality Management District, Diamond Bar, CA.

SCAQMD (South Coast Air Quality Management District). 2000b. MATES II. Multiple Air Toxics Exposure Study in the South Coast Air Basin. South Coast Air Quality Management District, Diamond Bar, CA [Online]. Available: http://www.aqmd.gov/matesiidf/matestoc.htm [Nov. 20, 2002].

Schauer, J.J., W.F. Rogge, M.A. Mazurek, L.M. Hildemann, G.R. Cass, and B.R. Simoneit. 1996. Source apportionment of airborne particulate matter using organic compounds as tracers. Atmos. Environ. 30(22):3837-3855.

Schwartz, J. 1999. Air pollution and hospital admissions for heart disease in eight U.S. counties. Epidemiology 10(1):17-22.

Scora, G., C. Levine, T. Younglove, and M.J. Barth. 2000. Second-by-Second Analysis of Catalyst and Fuel Enrichment Behavior in an In-Use Light-Duty Vehicle Fleet. Presentation at the 10th CRC On-Road Vehicle Emissions Workshop, March 27-29, 2000, San Diego, CA.

Seagrave, J.C., J.D. McDonald, A.P. Gigliotti, K.J. Nikula, S.K. Seilkop, M. Gurevich, and J.L. Mauderly. 2002. Mutagenicity and in vivo toxicity of combined particulate and semivolatile organic fractions of gasoline and diesel engine emissions. Toxicol. Sci. 70(2):212-226.

Segal, M., and R.A. Pielke. 1981. Numerical model simulation of human biometeorological heat load conditions - summer day case study for the Chesapeake Bay area. J. Appl. Meteor. 20:735-749.

Shephard, R.J. 1983. Carbon Monoxide, Silent Killer. Springfield, IL: Charles C. Thomas.

Shair, F.H., and K.L. Heitner. 1974. Theoretical model for relating indoor pollutant concentrations to those outside. Environ. Sci. Technol. 8(5):444-451.

Shelef, M. 1994. Unanticipated benefits of automotive emissions control: Reduction in fatalities by motor vehicle exhaust gas. Sci. Total Environ. 146/147:93-101.

Sheppard, L., D. Levy, G. Norris, T.V. Larson, and J.Q. Koenig. 1999. Effect of ambient air pollution on nonelderly asthma hospital admissions in Seattle, Washington, 1987-1994. Epidemiology 10(1):23-30.

Sheps, D.S., M.C. Herbst, A.L. Hinderliter, K.F. Adams, L.G. Ekelund, J.J. O'Neil, G.M. Goldstein, P.A. Bromberg, J.L. Dalton, M.N. Ballenger, S.M. Davis, and G.G. Koch. 1990. Production of arrhythmias by elevated carboxyhemoglobin in patients with coronary artery disease. Ann. Intern. Med. 113(5):343-351.

Sierra Research. 1999. Cold Start, Plug-in and Mid-Trip Idling CO Emissions from Light-Duty Gasoline-Powered Vehicles in Alaska. Report No. SR99-06-01. Prepared for the Alaska Department of Environmental Conservation, by Sierra Research, Inc., Sacramento, CA.

Singer, B.C., and R.A. Harley. 1996. A fuel-based motor vehicle emission inventory. J. Air Waste Manage. Assoc. 46(6):581-593.

Singer, B.C., and R.A. Harley. 2000. A fuel-based inventory of motor vehicle exhaust emissions in the Los Angeles area during summer 1997. Atmos. Environ. 34(11):1783-1795.

Smith, K.R. 2002. Place makes the poison: Wesolowski Award Lecture - 1999. J. Expo. Anal. Environ. Epidemiol. 12(3):167-171.

Smith, L., and B. Woodruff. 2001. Carbon Monoxide in Fort Collins. Presentation at the Third Meeting on Carbon Monoxide Episodes in Meteorological and Topographical Problem Areas, November 8, 2001, Denver, CO.

Stedman, D.H., G.A. Bishop, P. Aldrete, and R.S. Slott. 1997. On-road evaluation of an automobile emission test program. Environ. Sci. Technol. 31:927-931.

Stedman, D.H., G.A. Bishop, and R.S. Slott. 1998. The use of remote sensing measurements to evaluate control strategies: Measurements at the end of the first and second year of Colorado's biennial enhanced I/M program. Pp. 6.15-6.24 in Proceedings of the 8th CRC On-Road Vehicle Emissions Workshop, April 20-22, 1998, San Diego, CA, Vol.1. Coordinating Research Council, Inc., Atlanta, GA.

Therriault, S. 2002. Carbon Monoxide in Missoula. Presentation at the Sixth Meeting on Carbon Monoxide Episodes in Meteorological and Topographical Problem Areas, July 17, 2002, Missoula, MT.

TRB (Transportation Research Board). 2002. The Congestion Mitigation and Air Quality Improvement Program: Assessing 10 Years of Experience. Special Report 264. Washington, DC: National Academy Press.

Truex, T.J. 1999. Interaction of Sulfur with Automotive Catalysts and the Impact on Vehicle Emissions – A Review. SAE 1999-01-1543. Warrendale, PA: Society of Automotive Engineers.

U.S. Census Bureau. 2000a. United States Census 2000. Summary file 2 (SF2). U.S. Census Bureau, Washington, DC [Online]. Available: http://www.census.gov/Press-Release/www/2001/sumfile2.html [accessed April 30, 2003].

U.S. Census Bureau. 2000b. DP-1. Profile of General Demographic Characteristics: 2000 Data Set: Census 2000 Summary File 1 (SF1) 100-Percent Data, Geographic Area: Calexico City, California, and Fairbanks City, Alaska. [Online]. Available: http://www.census.gov. [accessed May 16, 2003].

U.S. Census Bureau. 2000c. (SU-99-3) Population Estimates for Cities with Populations of 10,000 and Greater (Sorted Within State by 1999 Population Size): July 1, 1999 (includes April 1, 1990 Population Estimates Base). Population Estimates Program, Population Division, U.S. Census Bureau, Washington, DC [Online]. Available: http://www.census.gov/population/estimates/metro-city/SC10K-T3.txt [accessed Nov. 20, 2002].

Verrelli, L. 2001. Early History of Carbon Monoxide Efforts in Anchorage and Fairbanks. Presentation at the First Meeting on Carbon Monoxide Episodes in Meteorological and Topographical Problem Areas, June 4, 2001, Washington, DC.

Wayson, R. L. 1999. Dispersion Modeling at Intersections: An Overview. Presented at Microscale Air Quality Impact Assessment for Transportation Projects, Transportation Research Board, Committee on Transportation and Air Quality (A1F03), January 10, 1999.

Whiteman, C.D., and J.C. Doran. 1993. The relationship between overlying synoptic-scale flows and winds within a valley. J. Appl. Meteorol. 32(11): 1669-1682.

Wilhelm, M., and B. Ritz. 2003. Residential proximity to traffic and adverse birth outcomes in Los Angeles county, California, 1994-1996. Environ. Health Perspect. 111(2):207-216.

Williams, M.L. 2000. Atmospheric pollution: Contribution of automobiles. Revue française d'allergologie et d'immunologie clinique 40(2):216-221.

Wrona, N. 1999. Questions Concerning the State's Remote Sensing Program. Draft. Memorandum to Herschella Horton, Assistant House Minority Leader, and John Loredo, House Minority Whip, from Nancy Wrona, Director, Air Quality Division, Arizona Department of Environmental Quality. Nov. 3, 1999.

Zhu, Y., W.C. Hinds, S. Kim, and C. Sioutas. 2002. Concentration and size distribution of ultrafine particles near a major highway. J. Air Waste Manage. Assoc. 52(9):1032-1042.

Glossary

ADEC. See *Alaska Department of Environmental Conservation.*
AFV. See *alternative fuel vehicle.*
Air-fuel ratio. The ratio, by weight, of air to gasoline entering the intake in a gasoline-powered engine. The ideal (stoichiometric) ratio for complete combustion is approximately 14.7 parts of air to 1 part of fuel, depending on the composition of the specific fuel.
Air quality model. A computer-based mathematical model used to predict air quality on the basis of emissions and the effects of the transport, dispersion, and transformation of compounds emitted into the air.
Air toxics. Air toxics refers to a host of carcinogens, respiratory toxicants, neurotoxicants, and other harmful atmospheric pollutants not included as criteria air pollutants. The Clean Air Act Amendments of 1990 listed 189 of these air toxics as hazardous air pollutants (HAPs) for future regulation. (One of the HAPS was removed in 1996, leaving 188 toxics on the HAP list.)
Alaska Department of Environmental Conservation (ADEC). The department that deals with clean air, land, and water issues in the state of Alaska
Albedo. The fraction of incoming sunlight that is reflected from the surface of the earth. Albedo is higher over whiter surfaces, such as snow cover, and lower over darker surfaces, such as oceans and forests.

Alternative fuel vehicle (AFV). Any dedicated, flexible-fuel, or dual-fuel vehicle designed to operate on at least one alternative fuel, such as compressed natural gas.
Ambient air. The air outside of structures. Often used interchangeably with "outdoor air."
Box model. A model that simulates how pollutant concentrations vary with time within a designated volume of air. The effects of emissions, chemical reactions, and exchange with the surrounding atmosphere are usually considered. The air is assumed to be well-mixed within the box.
CAA. See *Clean Air Act*.
CAAA90. See *Clean Air Act*.
California Air Resources Board (CARB). A part of the California Environmental Protection Agency whose mission it is to promote and protect public health, welfare, and ecological resources through the effective and efficient reduction of air pollutants, recognizing and considering the effects on the economy of the state.
CARB. See *California Air Resources Board*.
Carbon monoxide (CO). A colorless, odorless, tasteless, and toxic gas that results from the incomplete combustion of fuels containing carbon.
Carboxyhemoglobin (COHb). A molecule formed when CO reacts with hemoglobin, the intracellular protein that transports oxygen in the blood. The presence of carboxyhemoglobin increases hemoglobin's affinity for oxygen, thereby reducing the transport of oxygen from the blood to the body's tissues.
Clean Air Act (CAA). The original Clean Air Act was passed in 1963, but our national air pollution control program is actually based on the 1970 version of the law. The Clean Air Act Amendments of 1990 (CAAA90) are the most recent revisions of the law.
COHb. See *carboxyhemoglobin*.
Cold-start emissions. Tailpipe emissions that occur before a vehicle is fully warmed-up. Vehicle emissions are higher during the first few minutes of operation because the engine and the catalytic converter must come to operating temperature before they can become effective.
Conformity. See *transportation conformity*.
Criteria air pollutants. A group of six common air pollutants (*carbon monoxide*, lead, nitrogen dioxide, *ozone, particulate matter*, and sulfur dioxide) regulated by the federal government since the passage of the *Clean Air Act* in 1970, on the basis of information on health and/or environmental effects of each pollutant.

Diagnostic trouble codes. Codes that identify emissions control systems and/or components that are malfunctioning and are stored in the engine's computer. They can be retrieved using a *scan tool.*

Dose. The amount of a contaminant that is absorbed or deposited in the body of an exposed person for an interval of time—usually from a single medium. Total dose is the sum of doses received by interactions with all environmental media that contain the contaminant. Units (mass) of dose and total dose are often converted to units of mass or contaminant per volume of physiological fluid or mass of tissue.

Dynamometer. A treadmill-like machine that allows cars to be tested under the loads typical of on-road driving.

Emissions factor. See *emission rate.*

Emissions inventory. An estimate of the amount of a pollutant emitted into the atmosphere from major mobile, stationary, area-wide, and natural sources over a specific period of time, such as a day or a year.

Emissions rate. The amount of pollutant emissions produced by an activity per unit of activity. By using the emissions rate of a pollutant and data regarding quantities of materials used by a given source or its activity level, it is possible to compute emissions for the source. In the case of mobile-source emissions, estimated emissions are the product of an emissions rate in mass of pollutant per unit distance (e.g., grams per mile) and an activity estimate in distance (e.g., average miles traveled). In the case of stationary-source emissions, estimated emissions are the product of an emissions rate in mass of pollutant per unit energy (e.g., pounds per million Btus) and the amount of energy consumed.

Ethanol. Ethyl-alcohol is a volatile alcohol containing two carbon atoms (CH_3CH_2OH). For fuel use, ethanol is produced by fermentation of corn or other plant products.

Evaporative emissions. *Hydrocarbon* emissions that do not come from the tailpipe of a car but come from evaporation, permeation, seepage, and leaks in a car's fueling system. The term is sometimes used interchangeably with "nontailpipe emissions."

Exceedance. An air pollution event in which the ambient concentration of a pollutant exceeds a *National Ambient Air Quality Standard (NAAQS).*

Exceedance day. A day during which one or more exceedances takes place.

Exposure. An event that occurs when there is contact at a boundary between a human and an environmental contaminant of a specific concentration for an interval of time; the units of exposure are concentration multiplied by time.

Federal test procedure (FTP). A certification test for measuring the tailpipe and evaporative emissions from new vehicles over the urban dynamometer driving schedule, which attempts to simulate an urban driving cycle.

FTP. See *federal test procedure.*

Gaussian dispersion model. A microscale model that simulates the dispersion of pollution from a source such as an intersection or a factory. The spatial dispersion is assumed to be Gaussian in nature, and the ambient concentrations are assumed to be proportional to emissions.

Geostrophic winds. Large-scale winds controlled by pressure differences in the atmosphere. Geostrophic winds are not influenced much by the surface of the earth.

Gross vehicle weight rating (GVWR). The value specified by the manufacturer as the maximum design loaded weight of a single vehicle (i.e., vehicle weight plus rated cargo capacity).

GVWR. See *gross vehicle weight rating.*

HC. See *hydrocarbons.*

HDV. See *heavy-duty vehicles.*

HDDV. See *heavy-duty diesel vehicles.*

Heavy-duty diesel vehicles (HDDV). A *heavy-duty vehicle* using diesel fuel.

Heavy-duty vehicles (HDV). Any motor vehicle rated at more than 8,500 pounds *GVWR* or that has a vehicle curb weight of more than 6,000 pounds or a frontal area in excess of 45 square feet. This excludes vehicles that will be classified as medium-duty passenger vehicles for the purposes of the *Tier 2* emissions standards.

Hydrocarbons (HC). Organic compounds containing hydrogen and carbon.

I/M. See *inspection and maintenance.*

IM240. The name for the emissions test used in some I/M programs, including those in Arizona and Colorado. The IM240 is a transient, loaded-mode emissions test. "Loaded-mode" refers to the fact that the test is run on a treadmill-like device called a *dynamometer*, which simulates driving with the engine in gear. "Transient" refers to the fact that the car drives under a load that varies from second to second during the test. The "240" in IM240 indicates that the test lasts for 240 seconds. The shorter IM240 test is composed of pieces of *federal test procedure (FTP)* test activity and tends to correlate well with FTP results.

Inspection and maintenance (I/M). State emissions testing programs that attempt to identify vehicles with higher than allowable emissions and ensure that such vehicles are repaired or removed from the fleet.

Inversion. See *temperature inversion*.

Lapse rate. The rate at which temperature in the atmosphere changes with altitude. The average lapse rate is about $-6.5°C/km$. Under *temperature inversion* conditions, the lapse rate can be positive.

LDV. See *light-duty vehicle*.

LEV. See *low emission vehicle*.

Light-duty vehicle (LDV). A passenger car or passenger car derivative capable of seating 12 or fewer passengers. All vehicles and trucks under 8,500 GVWR are included (this limit previously was 6,000 pounds). Small pick-up trucks, vans, and sport utility vehicles (SUVs) may also be included.

Low-emission vehicle (LEV). A vehicle that meets EPA's Clean Fuel Vehicle or LEV standards or CARB's California LEV standards.

Malfunction indicator light (MIL). The instrument panel light used by *the onboard diagnostic (OBD) system* to notify the vehicle operator of an emissions related fault. The MIL is also known as the "service engine soon" or "check engine" lamp.

MIL. See *malfunction indicator light*.

Model year. Vehicles are certified for sale, marketed, and later registered as a certain "model year" indicating the year a vehicle was produced and offered for sale. Model years typically begin in September or October of the prior year, and run for roughly 12 months. In the last decade, certain vehicles have been introduced as a "pull-ahead" vehicle, appearing as early as January of the preceding year.

NAAQS. See *National Ambient Air Quality Standards*.

National Ambient Air Quality Standards (NAAQS). Standards set by EPA for the maximum levels of *criteria air pollutants* that can exist in the *ambient air* without unacceptable effects on human health or the public welfare.

Nitrogen oxides (NO_x). A general term referring to nitric oxide (NO) and nitrogen dioxide (NO_2). Nitrogen oxides are formed when air is raised to high temperatures, such as during combustion or lightning, and are major contributors to smog formation and acid deposition.

NO_x. See *nitrogen oxide*.

Nonattainment area. A geographic area in which the time-averaged concentrations of a *criteria air pollutant* have exceeded at some recent time a level allowed by the federal standards. A single geographic area may have acceptable levels of some criteria air pollutants but unacceptable levels of others; thus, an area can be both in attainment for one pollutant and in nonattainment for another.

Numerical predictive model. A mesoscale or large-scale model used to predict chemical concentrations in the atmosphere based on observed meteorological variables, emissions, and chemistry. Numerical predictive models represent the atmosphere as a three-dimensional grid of air parcels. Chemical transformations take place within the air parcels and air is transported between them.

O_3. See *ozone*.

OBD. See *onboard diagnostic system*.

OBDI. See *onboard diagnostics generation one*.

OBDII. See *onboard diagnostics generation two*.

Onboard diagnostic (OBD) systems. Devices incorporated into the computers of new motor vehicles to monitor the performance of the emission controls. The computer triggers a dashboard indicator light, referred to as a *malfunction indicator light (MIL)*, when the controls malfunction, alerting the driver to seek maintenance for the vehicle. The system also communicates its findings to repair technicians by means of diagnostics trouble codes, which can be downloaded from the vehicle's computer. Current OBD systems do not directly measure emissions.

Onboard diagnostics generation one (OBDI). An onboard automotive diagnostic system required by the *California Air Resources Board* since 1988 that uses a microprocessor and sensors to monitor and control various engine system functions. A *malfunction indicator light (MIL)* illuminates when a malfunction is noted, but engine technicians cannot connect to the system and download *diagnostic trouble codes*. (MIL flash patterns communicate the problem.)

Onboard diagnostics generation two (OBDII). OBDII expands upon OBDI to include monitoring of both the emissions system and sensor deterioration and to standardize the interface and code systems.

Oxides of nitrogen. See *nitrogen oxide*.

Oxyfuel. See *oxygenated fuel*.

Oxygen sensor. A sensor placed in the exhaust manifold to measure oxygen content. On some vehicles, oxygen sensors are located both before and after the catalytic converter.

Oxygenated fuel. Gasoline containing an oxygenate, typically methyl *tertiary*-butyl ether (MTBE) or ethanol, intended to reduce production of CO, a criteria air pollutant. In some parts of the country, CO emissions from cars make a major contribution to pollution. In some of these areas, gasoline refiners must market oxygenated fuels, which typically contain 2-3% oxygen by weight.

Oxygenates. Compounds containing oxygen (alcohols and ethers) that are added to gasoline to increase its oxygen content. Methyl *tertiary*-butyl ether (MTBE) and ethanol are the most common oxygenates currently used, although there are a number of others.

Ozone (O_3). A reactive gas whose molecules contain three oxygen atoms. It is a product of photochemical processes involving sunlight and ozone precursors, such as *hydrocarbons* and nitrogen oxides. Ozone exists in the upper atmosphere (stratospheric ozone), where it helps shield the earth from excessive ultraviolet rays, as well as in the lower atmosphere (tropospheric ozone) near the earth's surface. Tropospheric ozone causes plant damage and adverse health effects and is a *criteria air pollutant*; it is a major component of smog.

Particulate matter (PM). Any material, except uncombined water, that exists in the solid or liquid droplet states in the atmosphere. Particulate matter includes wind-blown dust particles, particles directly emitted as combustion products, and particles formed through secondary reactions in the atmosphere.

Photochemical reaction. A term referring to a chemical reaction brought about by sunlight, such as the formation of *ozone* from the interaction of *nitrogen oxides* and *hydrocarbons* in the presence of sunlight.

Plug-in. An electrical device used to heat the engine under extreme cold conditions in order to facilitate engine starting and reduce the time for emissions control devices to be activated.

PM. See *particulate matter*.

$PM_{2.5}$. A subset of particulate matter that includes those particles with an aerodynamic equivalent diameter less than or equal to a nominal 2.5 micrometers (μm). This fraction of PM penetrates most deeply into the lungs, and causes the majority of visibility reduction.

PM_{10}. A subset of particulate matter that includes those particles with an aerodynamic equivalent diameter less than or equal to a nominal 10 μm (about 1/7 the diameter of a single human hair). This fraction of PM causes visibility reduction and can penetrate into the lungs.

Preconditioning. A set of steps followed to warm-up a vehicle prior to an I/M emissions test so that it can give valid results. Cut points, which determine passing or failing for a vehicle, are based on testing a fully warmed-up vehicle in which the emissions control equipment, including the catalytic converter, are hot and fully functional. If an owner drives a short distance to the test station or if the vehicle has to wait in the test station for a long time, it may not be fully warmed up. This may result in a false reading; a car that would have passed if fully warmed (i.e., fully preconditioned) might fail.

Process numerical model. A mesoscale or large-scale model used to analyze atmospheric processes and their impacts on air quality, typically for research purposes. Like *numerical predictive models*, process numerical models represent the atmosphere as a three-dimensional grid of air parcels. Process numerical models better resolve the coupling between meteorology and chemistry than numerical predictive models and are not necessarily constrained by observations.

Reformulated gasoline (RFG). Specifically formulated fuels blended such that, on average, the exhaust and evaporative emissions of VOCs and hazardous air pollutants (chiefly benzene, 1,3-butadiene, polycyclic aromatic hydrocarbons, formaldehyde, and acetaldehyde) are significantly and consistently lower than such emissions resulting from use of conventional gasolines. The 1990 Clean Air Act amendments required sale of reformulated gasoline in the nine areas with the most severe ozone pollution problems. RFG contains, on average, a minimum of 2.0 weight percent oxygen.

Remote sensing. A method for measuring pollutant concentrations from a vehicle's exhaust while the vehicle is traveling down the road. Remote-sensing systems employ infrared absorption to measure VOC and CO concentrations relative to carbon dioxide. These systems typically operate by continuously projecting a beam of infrared radiation across a roadway and making measurements on the exhaust plume after a vehicle passes through the beam.

RFG. See *reformulated gasoline*.

Scan tool. A hand-held computer that is plugged into a vehicle's OBD data link connector to allow a technician to read *diagnostic trouble codes*, readiness status, and monitor other information collected by the OBD system.

Secondary particulate matter. *Particulate matter* that is formed in the atmosphere, and is generally composed of species such as ammonia or the products of atmospheric chemical reactions, such as nitrates, sulfates and organic material, in addition to some water. Secondary particles are distinguished from primary particles, which are emitted directly into the atmosphere.

SIP. See *state implementation plan*.

State implementation plan (SIP). A detailed description of the scientific methods and programs a state will use to carry out its responsibilities under the *Clean Air Act* for complying with the NAAQS. SIPs are a collection of the programs used by a state to reduce air pollution. The Clean Air Act requires that EPA approve each SIP, after the public has had an opportunity to participate in its review and approval.

Statistical roll-back model. A model that estimates the emission reductions needed for a desired improvement in air quality. The emission reductions are assumed to be linearly related to ambient concentrations of pollutants.

Subsidence. Slow descent of air cooled by radiative cooling.

Super ultra-low emission vehicle (SULEV). A vehicle that produces fewer exhaust emissions than *ultra-low emission vehicles (ULEV)*.

Supplemental federal test procedure (SFTP). The SFTP is a certification test for measuring the tailpipe and evaporative emissions from new vehicles that includes two driving cycles not represented in the FTP. The SFTP includes a test cycle simulating high speed and high acceleration driving (US06 cycle) and a test cycle that evaluates the effects of simulating air conditioner operation (SC03 cycle).

Synoptic. Used to describe meteorological processes that occur over regional spatial scales over several days.

TCM. See *transportation control measure*.

TDM. See *transportation-demand management strategies*.

Temperature inversion. An atmospheric condition in which temperature in the lower part of the atmosphere increases with altitude, rather than decreasing with altitude, as is more typical. Inversion conditions can trap pollution near the surface because warmer, less dense air is resting above colder, more dense air.

Three-way catalytic converter. A catalytic converter designed to both oxidize CO and VOCs and reduce NO_x emitted from gasoline-fueled vehicles.

Tier 0 vehicles. Vehicles that meet Tier 0 tailpipe standards. For *light-duty vehicles*, these standards began with *model year* 1981 vehicles and were phased out in model year 1995 for passenger cars and most light-duty trucks.

Tier 1 vehicles. Vehicles that meet Tier 1 tailpipe standards. For *light-duty vehicles*, these standards began with *model year* 1994 vehicles.

Tier 2 vehicles. Vehicles that will meet Tier 2 tailpipe standards. For *light-duty vehicles*, these standards would not begin until *model year* 2004 vehicles.

Transitional low emission vehicle (TLEV). A vehicle meeting either EPA's Clean Fuel Vehicle TLEV standards or CARB's California Low-Emission Vehicle Program TLEV standards. TLEVs produce fewer emissions than federal *Tier 1 vehicles*.

Transportation conformity. A process to demonstrate whether a federally supported activity is consistent with the air quality goals in *state implementation plans (SIPs)*. Transportation conformity demonstrates that plans, programs, and projects approved or funded by the Federal Highway Administration or the Federal Transit Administration for regionally significant projects do not create new violations, increase the frequency or severity of existing violations, or delay timely attainment of the *National Ambient Air Quality Standards (NAAQS)*.

Transportation control measure (TCM). Any control measure to reduce vehicle trips, vehicle use, vehicle-miles traveled, vehicle idling, or traffic congestion for the purpose of reducing motor-vehicle emissions. TCMs can include encouraging the use of carpools and mass transit.

Transportation-demand management (TDM) strategies. Strategies which use regulatory mandates, economic incentives, or educational campaigns to change driver behavior. TDM strategies attempt to reduce the frequency or length of automobile trips or to shift the timing of automobile trips.

Transportation plan. A long-range plan that identifies facilities that should function as an integrated transportation system. Under the Intermodal Surface Transportation Efficiency Act of 1991, metropolitan planning organizations (MPOs) must have transportation plans in place that present a 20-year perspective on transportation investments for their region. The transportation plan gives emphasis to those facilities that serve important national and regional transportation functions, and includes a financial plan that demonstrates how the long-range plan can be implemented.

Transportation-supply improvement (TSI) strategies. TSI strategies attempt to reduce emissions by changing the physical infrastructure of road system to improve traffic flow and reduce stop-and-go movements.

Travel-demand model. An analysis procedure using heuristics or formal systems of equations to estimate the number, distribution, mode choice, and/or route choice of trips made by a household or individual that can be aggregated to estimate the number of trips starting and/or ending in a specific geographical area. The model determines the amount of transportation activity occurring in a region based on an understanding of the daily activities of individuals and employers, as well as the resources and transportation infrastructure available to households and individuals when making their daily activity and travel decisions.

TSI. See *transportation-supply improvement strategies*.

Two-way catalytic converter. A first generation catalytic converter designed to oxidize CO and VOC emissions from gasoline-fueled vehicles.

UAM. See *urban airshed model*.

ULEV. See *ultra-low emission vehicle*.

Ultra-low emission vehicle (ULEV). A vehicle meeting either EPA's Clean Fuel Vehicle ULEV standards or CARB's California Low-Emission Vehicle Program ULEV standards. ULEVs produce fewer emissions than LEVs.

Urban airshed model (UAM). A three-dimensional photochemical air quality grid model for calculating the concentrations of both inert and chemically reactive pollutants in the atmosphere. It simulates the physical and chemical processes that affect concentrations of pollutants. The UAM was a specific model developed by Systems Application International, Inc., but the term is now often used generically to describe a variety of models used in this field.

Vehicle-miles traveled (VMT). The number of miles driven by a fleet of vehicles over a set period of time, such as a day, month, or year. One vehicle traveling one mile is one vehicle-mile.

VMT. See *vehicle-miles traveled*.

VOC. See *volatile organic compounds*.

Volatile organic compounds (VOCs). Organic compounds that can include oxygen-, nitrogen-, and sulfur-containing compounds. Alkanes, alkenes and aromatic hydrocarbons are all VOCs (as well as being HCs). The simple carbon-containing compounds CO and carbon dioxide are usually classified as inorganic compounds. A volatile organic compound is one that can exist as a gas at ambient temperatures. Many volatile organic chemicals are hazardous air pollutants; for example, benzene causes cancer.

Zero emission vehicle. A vehicle that emits no tailpipe exhaust emissions.

Appendix A

Biographical Information on the Committee on Carbon Monoxide Episodes in Meteorological and Topographical Problem Areas

Armistead G. Russell (*Chair*) is the Georgia Power Distinguished Professor of Environmental Engineering at the Georgia Institute of Technology. His research areas include air pollution control, aerosol dynamics, atmospheric chemistry, emissions control, air pollution control strategy design and computer modeling. Dr. Russell has served on a number of NRC committees and was chair of the Committee to Review EPA's Mobile Source Emissions Factor (MOBILE) Model. He received a Ph.D. in mechanical engineering from the California Institute of Technology.

Roger Atkinson is a research chemist and Distinguished Professor of Atmospheric Chemistry at the University of California at Riverside. His research areas include the kinetics and mechanisms of atmospherically important reactions of organic compounds in the gas phase. Dr. Atkinson serves on the California Air Resources Board's Reactivity Scientific Advisory Committee and the California Air Resources Board's Scientific Review Panel on Air Toxics, and has served on NRC committees including the Committee on Tropospheric Ozone Formation and Measurement and the Committee on Ozone-Forming Potential of Reformulated Gasoline. He received a Ph.D in physical chemistry from the University of Cambridge.

Sue Ann Bowling is a retired professor from the University of Alaska, Fairbanks. Her research interests included air pollution meteorology, polar meteorology, radiative transfer, paleoclimatology, and climatic change. Dr. Bowling received a Ph.D. from the University of Alaska, Fairbanks.

Steven D. Colome is deputy director of the Southern California Particle Center and Supersite and an adjunct professor in environmental health at the University of California, Los Angeles School of Public Health. His research interests include human exposure assessment, environmental epidemiology, indoor air quality, regional exposure modeling, and health effects assessment. Dr. Colome previously served on the NRC Committee on Toxicological and Performance Aspects of Oxygenated Motor Vehicle Fuels. He was a reviewer of the EPA document Air Quality Criteria for Carbon Monoxide. He received a Sc.D. in environmental health sciences from Harvard University, School of Public Health.

Naihua Duan is professor in residence in the Department of Psychiatry and Biobehavioral Sciences and the Department of Biostatistics at the University of California, Los Angeles. Previously he was a corporate chair and Senior RAND Fellow in statistics at RAND. His research interests include nonparametric and semiparametric regression methods, sample design, hierarchical models, and environmental exposure assessment, including exposure to carbon monoxide. He served as a member of the NRC Committee on Advances in Assessing Human Exposure to Airborne Pollutants. Dr. Duan received a Ph.D. in statistics from Stanford University.

Gerald Gallagher is president of J Gallagher and Associates. Previously, he served as manager of the Mobile Sources Program for the Air Pollution Control Division of the Colorado Department of Public Health and Environment. His responsibilities included the development and implementation of air quality management plans for controlling carbon monoxide. He was also responsible for the operation of a metro-wide inspection and maintenance program, consisting of approximately 1.8 million inspections per year for gasoline- and diesel-fueled vehicles. Dr. Gallagher is a member of the NRC Committee on Vehicle Emissions Inspection and Maintenance Programs. He received a Ph.D. in intergovernmental relations/environmental management from the University of Colorado.

Randall L. Guensler is an associate professor in the School of Civil and Environmental Engineering at Georgia Institute of Technology. His research interests include the relationships between land use, infrastructure, travel behavior, and vehicle emission rates; transportation and air quality planning and modeling—theory and practice; and emission control strategy effectiveness. Dr. Guensler is the former chairman of the Transportation Research Board's Committee on Transportation and Air Quality. He received a Ph.D. in civil and environmental engineering from the University of California, Davis.

Susan L. Handy is an associate professor in the Department of Environmental Science and Policy at the University of California at Davis. Her research interests focus on the relationship between transportation systems and land use patterns, and the effects of telecommunications technologies on patterns of development and travel behavior. Dr. Handy is currently chair of the Transportation Research Board Committee on Telecommunications and Travel Behavior, and also serves on the Committee on Transportation and Land Development. She received a Ph.D. in city and regional planning from the University of California, Berkeley.

Simone Hochgreb is professor of experimental combustion at the University of Cambridge in England. Her research focuses on fundamental and applied problems in combustion and chemical kinetics, with particular focus on applications to transportation, internal-combustion engines, and pollutant emission formation. Dr. Hochgreb served as a member of the NRC Committee on Toxicological and Performance Aspects of Oxygenated Motor Vehicle Fuels as well as on the NRC Review Panel for the Partnership for a New Generation of Vehicles. She received a Ph.D. in mechanical and aerospace engineering from Princeton University.

Sandra N. Mohr is a consultant. Dr. Mohr's research interests focus on the health effects of air pollutants. She has been a lead researcher in the health effects of methyl *tertiary*-butyl ether (MTBE), a gasoline additive, and has served on the NRC's Committee on Toxicology and Performance Aspects of Oxygenated Motor Vehicle Fuels. She received an M.D. from the University of Kansas School of Medicine and the M.P.H. degree from Yale University.

Roger A. Pielke Sr. is a professor in the Department of Atmospheric Science at Colorado State University. He is also State Climatologist for Colorado. His research areas include the study of global, regional, and local weather and climate phenomena through the use of sophisticated mathematical simulation models, air pollution meteorology, and mesoscale meteorology. Dr. Pielke Sr. received a Ph.D. in meteorology from Pennsylvania State University.

Karl J. Springer is retired Vice President for Automotive Products and Emissions Research at Southwest Research Institute. His research interests have focused on the measurement and control of air pollution emissions from on-road and off-road vehicles and equipment powered by internal combustion engines. Mr. Springer is a member of the National Academy of Engineering. He received a BSME from Texas A & M and an M.S. in physics from Trinity University.

Roger Wayson is a professor of civil and environmental engineering at the University of Central Florida where he conducts research in the microscale modeling of carbon monoxide ambient concentrations that result from mobile sources and airport operations. Dr. Wayson obtained his B.S. and M.S. in environmental engineering from the University of Texas at Austin and his Ph.D. in civil engineering from Vanderbilt University.

Appendix B

Abbreviations and Names Used for Classifying Organic Compounds (NRC 1999)

VOC[1] (volatile organic compound)—Organic compounds that are found in the gas phase at ambient conditions. Might not include methane.

ROG (reactive organic gas)—Organic compounds that are assumed to be reactive at urban (and possibly regional) scales. Definitionally, taken as those organic compounds that are regulated because they lead to ozone formation. Does not include methane. The term is used predominantly in California.

NMHC (nonmethane hydrocarbon)—All hydrocarbons except methane; sometimes used to denote ROG.

NMOC (nonmethane organic compound)—Organic compounds other than methane.

RHC (reactive hydrocarbon)—All reactive hydrocarbons; also used to denote ROG.

THC (total hydrocarbon)—All hydrocarbons, sometimes used to denote VOC.

OMHCE (organic material hydrocarbon equivalent)—Organic compound mass minus oxygen mass.

TOG (total organic gas)—Used interchangeably with VOC.

---

[1] Unless noted otherwise, HC is the term used in this report to represent the general class of gaseous organic compounds.

Appendix C

A Simple Box Model with Recirculation

One can formulate a simple model to follow the evolution of pollutant concentrations based on looking at the air mass above an area of interest as a well-mixed box. In this case, the time varying volume of the box, $V(t)$, is given as the area over which the model is being applied, multiplied by the effective height into which the pollutants are mixed (i.e., the mixing height, $H(t)$), noting that it can vary with time.

$$V(t) = LWH(t)$$

where L is the length of the box (in the direction of the wind) and W is the width of the box perpendicular to the wind, as shown in Figure C-1.

Applying the principle of conservation of mass, one then gets

$$\frac{dLWHc}{dt} = M_{in} - M_{out} + E$$

where M_{in} is the mass flux in (mass/time) due to the wind carrying the contaminant (CO) in from outside and M_{out} is the mass flux out; c is the time dependent CO concentration (mass/volume) in the box, and E the time dependent emission rate from all sources in the box.

The mass flux in is given by the wind velocity, U, multiplied by the area of the side of the box (WH) and background CO concentration c_b,

$$M_{in} = UWH(t)c_b.$$

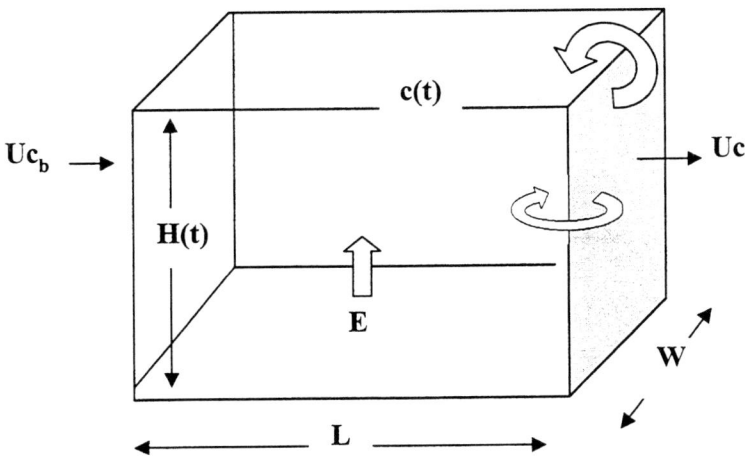

FIGURE C-1 Diagram of a simple box model with a box of width W, length L, and time dependent height $H(t)$ and wind speed U. E is the total mass rate of pollutant emissions within the box, which is assumed to be well mixed, with CO concentration c throughout; c_b is the background concentration.

If the background concentration is small it can be neglected. As noted below, the assumption that the mass flow into the box depends only on the background concentration may be wrong if there is CO that leaves the box but is recirculated. The mass flux out is given by

$$M_{out} = UWH(t)c$$

where the assumption that the box is well mixed leads to having the contaminant concentration leaving the box the same as that within it. Using the above two expressions for M_{in} and M_{out} leads to the classical box model formulation for the evolution of pollutant concentration c,

$$\frac{\Delta c}{\Delta t} = \frac{Uc_b}{L} - \frac{Uc}{L} + \frac{E}{LWH(t)} - \frac{1}{H(t)}\frac{dH}{dt}\bigg|_{\frac{dH}{dt}>0}.$$

The last term causes the concentration to decrease if the mixing height increases with time, because the CO mixes in an expanding volume. If the mixing height decreases with time, the change in height has no effect on the concentration in the box (the last term is zero).

If the emission rate is much greater than the flux in due to the wind ($E \gg M_{in}$), the first term can be dropped. This equation can account for the time variation in both the mixing height and emission rate, but assumes that the length and width of the box are fixed. Further, it assumes that the pollutant concentrations do not vary spatially within the box, which implies that the emissions do not vary significantly spatially. This limits the size of the model application area, and means that it should not be used to estimate hot spot concentrations. Also, it assumes that none of the contaminant that leaves the box returns, except after being added to the background levels. This assumption can be wrong. In that case, the model should be modified to account for the fraction of contaminant that leaves the box and is recirculated back into it. This can be done by adding a recirculation coefficient α that is the fraction of contaminant that returns to the box after being advected out. Physically, α must be between 0 and 1. The resulting model becomes

$$\frac{\Delta c}{\Delta t} = \frac{Uc_b}{L} - \frac{U(1-\alpha)c}{L} + \frac{E}{LWH(t)} - \frac{c}{H(t)}\frac{dH}{dt}\bigg|_{\frac{dH}{dt}>0} = \frac{Uc_b}{L} + \frac{E}{LWH(t)} - \left[\frac{U(1-\alpha)}{L} + \frac{1}{H(t)}\frac{dH}{dt}\bigg|_{\frac{dH}{dt}>0}\right]c.$$

If α is zero, the original box model is obtained. As α approaches one, virtually all of the air and contaminant leaving the box re-enters, allowing the contaminant concentration to build up dramatically.